Einstein on Race and Racism

FRED JEROME AND RODGER TAYLOR

Rutgers University Press
New Brunswick, New Jersey, and London

First paperback printing, 2006

Library of Congress Cataloging-in-Publication Data

Jerome, Fred.
Einstein on race and racism / Fred Jerome and Rodger Taylor.
p. cm.
Includes bibliographical references and index.
ISBN 0–8135–3617–0 (hardcover : alk. paper)
ISBN 978–0–8135–3952–2 (pbk. : alk. paper)
1. Einstein, Albert, 1879–1955—Views on racism. I. Taylor, Rodger, 1953– II. Title.
QC16.E5J466 2005
530′.092—dc22 2004023485

A British Cataloging-in-Publication record for this book is available from the British Library.

Manufactured in the United States of America

To the people of Witherspoon Street,
yesterday and today

Contents

Preface

More than one hundred biographies and monographs about Albert Einstein have been published, yet not one of them mentions the name Paul Robeson, let alone Einstein's friendship with him; or the name W.E.B. Du Bois, let alone Einstein's support for him. Nor does one find in any of these works any reference to the Civil Rights Congress whose campaigns Einstein actively supported. Finally, nowhere in all the ocean of published Einsteiniana—anthologies, bibliographies, biographies, summaries, articles, videotapes, calendars, posters and postcards—will one find even an islet of information about Einstein's visits and ties to the people in Princeton's African American community around the street called Witherspoon.[1]

One explanation for this historical amnesia is that Einstein's biographers and others who shape our official memories felt that some of his "controversial" friends, such as Robeson, and activities, such as co-chairing the antilynching campaign, might somehow tarnish Einstein as an American icon. That icon, sanctified by *Time* magazine when it dubbed Einstein the "Person of the Century," is a myth, albeit a marvelous myth. In fact, as myths go, Einstein's is hard to beat. The world's most brilliant scientist is also a kindly, lovably bumbling, grandfather figure: Professor Genius combined with Dr. Feelgood! Opinion-molders, looking down from their ivory towers, may have concluded that such an appealing icon will help the great unwashed public feel good about science, about history, about America. Why spoil such a beautiful image with stories about racism, or for that matter with any of Einstein's political activism? Politics, they argue, is ugly, making teeth grind and fists clench, so why splash politics over Einstein's icon? Why drag a somber rain cloud across a bright blue sky? Einstein might reply, with a wink, that without rain clouds life would be very, very short. Or he might simply say that a bright blue sky is a fairy tale in today's war-weary world.

Yet, despite Einstein's clear intention to make his politics public—especially his antilynching and other antiracist activities—the history-molders have seemed embarrassed to do so. Or nervous. "I had to think about my Board," a museum curator (who doesn't want his name used even today) stated, while explaining his omission of some of the scientist's political statements from the major exhibition celebrating Einstein's one-hundredth birthday in 1979.[2]

When it came to how to handle Einstein's ashes or his house on Mercer Street, everyone involved meticulously adhered to his wishes. But when it involved his ideas, and especially his concerns about what he called America's "worst disease," the fact that Einstein wanted his views made as public as possible seems to have slipped past his historians.

Readers may judge for themselves how much of this oversight is due to forgetting and how much may be due to other motives (including, perhaps, disagreement with Einstein's point of view). It is not so much the motive for the omission but the consequence that concerns us. Americans and the millions of Einstein fans around the world are left unaware that Einstein was an outspoken, passionate, committed antiracist. "It is certain—indeed painfully obvious—that racism has permeated U.S. history both as idea and practice," as the historian Herbert Aptheker states. Nevertheless, he adds, "It always has faced significant challenge."[3]

Racism in America depends for its survival in large part on the smothering of antiracist voices, especially when those voices come from popular and widely respected individuals—like Albert Einstein. This book, then, aspires to be part of a grand un-smothering.

Acknowledgments

Our thanks to Meg Rich for invaluable guidance at the Special Collections in Princeton University's Firestone Library, as well as Annalee Pauls and Charles Greene; to Dan Bauer, whose memory and files provided information on the history of McCarter Theatre programs; and Maureen M. Smyth, who led us through the files of the Princeton Historical Society with verve and understanding. Our gratitude also to Daniel J. Linke, archivist and curator of Public Policy Papers at Seeley G. Mudd Manuscript Library, Princeton University, his assistant, Tad Bennicoff, and the entire superb staff at the Mudd Library; to Joey Cotruvo and Donald Marsden of Princeton University's Triangle Club; to Susan Pevar and the other good folks at Lincoln University, Pennsylvania, for their generous assistance and hospitality and to Yvonne Foster Southerland for sharing her recollections and photos of Einstein's visit to Lincoln; to Terri Nelson at the Princeton Public Library and Chris Moore at the Schomburg Center for Research in Black Culture, New York Public Library, for their knowledge and insight. Super-helpful, too, were the Schomberg Center's entire librarian staff and Barbara Wolff, Chaya Becker, and Dr. Roni Grosz at the Einstein Archives at Hebrew University in Jerusalem. Thanks also to Peter Filardo, archivist, and Dr. Gail Malmgreen, associate head for archival collections, at NYU's Tamiment Library; and to Julia Newman, executive director of the Abraham Lincoln Brigade Archives (ALBA).

To all the folks in the Witherspoon community, with extra special thanks to Alice and Shirley Satterfield for making their home available and inviting others to two wonderful lunch-gatherings; and to Hank Pannell for the extra outreach in contacting folks, taking photos, and just being there always with can-do reassurance. To Mr. Albert Hinds for sharing his unfaded 102-year-old memory with us; to Fanny and Jim Floyd for spending extra time and effort teaching us about the Witherspoon community and its history; to Mary Trotman for taking time from her overburdened work schedule and helping us to meet her mother, Lillie, and Lillie's granddaughter Lena.

To Jack Washington for sharing his valuable time and historical knowledge; to Emily Belcher; and to Nell Painter and Valerie Smith in Princeton University's Program in African American Studies.

To our friends and families who helped beyond help in reading, criticizing, encouraging, and, most of all, tolerating our efforts: To Debbie, Jocelyn, Romare, Rebecka, Mark, and Daniel—with double-duty gratitude to Jocelyn for lending her insightful and persistent research and to Rebecka for skillful translation assistance—to Cheryl Gross, Michelle Boyd, James Boyd, Ernie Herman, and Bob Apter; to Nils and Gunilla Smith-Solbakken for providing "a clean, well-lighted place"; and to Florence Taylor for seeing Einstein walk through Princeton.

Special thanks to John Stachel and Dorothy Zellner for meticulously and caringly correcting errors in our early drafts. And especially to Kitsi Watterson, not only for so unselfishly sharing her research and introducing us to so many friends, but also for her encouragement from the start—"you simply must do this project." To our ever-young agent Frances Goldin for her constant support; to Audra Wolfe, our diligent editor at Rutgers University Press; and, last but furthest from least, to Michael Denneny, whose editorial assistance was magical.

PART I

EINSTEIN AND ROBESON ON WITHERSPOON STREET

Your ancestors dragged these black people from their homes by force; and in the white man's quest for wealth and an easy life they have been ruthlessly suppressed and exploited, degraded into slavery. The modern prejudice against Negroes is the result of the desire to maintain this unworthy condition. . . . I believe that whoever tries to think things through honestly will soon recognize how unworthy and even fatal is the traditional bias against Negroes.

—Albert Einstein

CHAPTER 1

Escape from Berlin

On January 30, 1933, the day Hitler and the Nazis took over the German government, the most famous scientist in the world may also have been the luckiest. Albert Einstein and his wife, Elsa, were away from their Berlin home, on a visit to Pasadena, California—his third winter there as a guest faculty member at the California Institute of Technology. The Einsteins had planned to return home in the spring, but that was before January 30. Within a few months, the Nazi regime made it clear that Einstein was still alive primarily because he was not in Germany.

Einstein, more than any other scientist, arguably more than any other human being, by his very existence—a genius who was also a Jew, a democrat, and, later, a socialist—gave the lie to Hitler's Nazi theories.

Even before the Nazis started calling themselves Nazis (before Hitler's National Socialist party emerged in the mid-1920s), right-wing German nationalists had targeted Einstein for attack:

> Some of these Nationalists took to waiting for Einstein outside his apartment on Haberlandstrasse, or his office in the Prussian Academy of Science, and shouting denunciations of "Jewish science" as soon as the familiar figure appeared. Others filled his mailbox with obscene and threatening letters. On one occasion, a gang of right-wing students disrupted Einstein's lecture at Berlin University and one of them shouted, "I'm going to cut the throat of that dirty Jew." An anti-Semitic demagogue named Rudolph Leibus

> was arrested [in 1921]—and fined the trivial sum of sixteen dollars—for offering a reward to anyone who would assassinate the hated scientist.[1]

And while he was invited to speak and hailed by audiences around the world—one trip took the Einsteins to China, Japan, Palestine, and Spain, where they were cheered by hundreds of thousands—in Germany, a group calling itself the Committee of German Scientists for the Preservation of Pure Scholarship launched an attack on Einstein, labeling the theory of relativity "a Jewish perversion." Even in winning the Nobel Prize in physics for 1921, Einstein faced anti-Semitism.*

For several years in the mid-1920s, *public* anti-Semitism subsided and the ultra-nationalists lowered their profile as the German economy steadied with substantial economic aid from the United States and England. But toward the end of the decade, the economy faltered, and the far right flexed its political and paramilitary muscles. Hitler's Aryan-supremacy racism infected millions of Germans seeking scapegoats for their economic difficulties and loss of international influence. And when the worldwide economic depression of 1929 brought rampant unemployment and inflation to Germany, it also brought the Nazis a large, disgruntled base of potential recruits.

The Nazi party, led by Hitler, had first made headlines in 1923 with its "Beer Hall Putsch," a failed attempt at taking over the government in the German state of Bavaria. Following the "Putsch," the party grew slowly, and by 1929 it had only 12 representatives in the Reichstag (the German Congress). The im-

*Einstein had been nominated for the Nobel Prize in ten of the previous twelve years (1909–1920), yet it wasn't until he had been hailed as a world-renowned celebrity that the Nobel committee agreed to award him their prize. Years later, Irving Wallace, author of *The Prize*, interviewed Sven Hedin, one of the Nobel judges, who acknowledged that anti-Semitism had influenced the judges to vote repeatedly against an award for Einstein (Wallace, *The Writing of One Novel*). Hedin later publicly supported the Nazis and was a friend of Göring, Himmler, and Hitler.

pact of the economic depression and the success of Nazi propaganda brought a surge of Nazi votes in the 1930 elections, increasing their Reichstag seats to 107.

Nazi street gangs launched increasingly violent attacks against Hitler's enemies, especially leftists and Jews. As one historian put it, "To ready the ground for the Nazi rise to ultimate power, the party raised the level of violence witnessed by ordinary Germans with each passing month."

> An incident on June 10, 1932, was typical of the strategy. That afternoon, several hundred members of the Nazi SA and . . . SS private armies invaded the working-class district of Berlin-Wedding . . . blocked the ends of a stretch of road . . . chanting anti-Semitic slogans and . . . attacking anyone luckless enough to be out and about. The Nazis beat up some thirty locals, including several old people and one pregnant woman, who was hospitalized in dire condition.[2]

There was no question about the identity of the Nazis' number-one target. Their hostility, in the words of Einstein biographer Philipp Frank, "was concentrated to an astonishing and . . . frightening degree against Einstein."[3] In 1929, a Leipzig publishing house issued a book titled *One Hundred Authors against Einstein*. The book itself had little impact, "but it was a warning." Attacks against prominent Jews may have faded during the stabilized mid-1920s, "but now . . . the threat was back."[4] A friend who visited Einstein in Germany in 1930 described the growing signs of anti-Semitism—"Many Jewish shops had been sacked"—and reported that the scientist, "for all his serenity, was anxious."[5]

The threats against Einstein increased as Hitler came closer to power. A local baker in Caputh, the village near Potsdam where the Einsteins had built a summer house, began to complain loudly to his customers about the scientist's "Jewish house." In late spring 1932, the scientist stopped walking alone, and Elsa's friend Antonina Vallentin warned her that to "leave Einstein in

Germany was to perpetrate a murder."[6] Just before he and Elsa left Germany for the last time in December 1932, he received a "friendly warning" from a top German general that his life "is not safe here anymore."[7]

Officially, the Einsteins were departing for one more semester abroad and planned to return to Berlin in the spring. Einstein told the *New York Times,* "I am not abandoning Germany. . . . My permanent home will still be in Berlin."[8] But he may have had a suspicion they would not return: when the steamer *Oakland* left Bremerhaven on December 10, 1932, it carried the Einsteins and thirty pieces of luggage. It was, as Einstein's friend and biographer Abraham Pais put it, "a little excessive for a three months' absence."[9]

In America, Einstein was quickly vilified by the German state. He was assailed as the chief of a secret anti-Nazi movement, sometimes described as "communistic," sometimes as the "Jewish International."[10]

On March 23, the Third Reich barred Jews and Communists from teaching in universities and working as lawyers or in civil service jobs. Scientists, especially Jewish scientists, were a special target for the regime that preached Aryan supremacy. One Nazi pedagogical leader put it plainly: "It is not science that must be restricted, but rather the scientific investigators and teachers; only . . . men who have pledged their entire personality to the nation, to the racial conception of the world . . . will teach and carry on research at the German universities."[11]

The Nazis repeatedly raided the Einsteins' Berlin apartment, seized all their belongings, and froze their bank account. In March, Nazi SA agents ransacked their summer house in Caputh, searching for a secret cache of weapons "allegedly hidden there by Communists" and intended for an anti-Nazi uprising. When they found no weapons—"all they found was a bread knife," the *New York Times* reported—they confiscated the house anyway, declaring it was "obviously" about to be sold to finance subversive activities.

Einstein was suddenly a refugee. Even if he might, miraculously, have survived a return to Germany, he told the press: "As long as I have any choice in the matter, I shall live only in a country where civil liberty, tolerance and equality of all citizens before the law prevail. . . . These conditions do not exist in Germany at the present time."[12] But the Einsteins did return to Europe in the spring and summer of 1933, spending several months in the Belgian coastal town of Le Coq sur Mer for rest and reconsideration of future plans. On hearing that Nazi newspapers had put a price of $5,000 on his head, Einstein quipped, "I didn't know I was worth so much." Nonetheless, the death threats were serious: during his stay at Le Coq, the Belgian government assigned two twenty-four-hour bodyguards to protect him from a reported Nazi assassination team. When the Hitler regime issued an official book of photos of "Enemies of the State," the caption under Einstein's photo read *Noch Ungehängt* (Not Yet Hanged).[13]

Einstein was also wanted—but alive and thinking—by leading institutions of learning around the world. Several European universities, including those in Oxford, Paris, Madrid, and Leiden, offered Einstein faculty positions, as did the newly established—and well-funded—Institute for Advanced Study in Princeton.* Einstein felt at home in Europe, but, as author and physicist C. P. Snow explains, the choice of where to settle was, to a large extent, out of his hands:

*Designed from the start as a center exclusively for theoretical research, the Institute, and its founder, Abraham Flexner, hoped to attract the most outstanding scholars from around the world to live, think, and do research at the IAS, which, although located in Princeton, would be independent of the university. Besides their obvious differences—the Institute was not a degree-granting institution and had no students—one other key policy difference central to this story is that the IAS, unlike Princeton University, welcomed Jewish scholars from around the world. To a large extent, the Institute reflected America's emerging status as the leading financial and technological power in the world. Einstein was the Institute's most valuable asset, ensuring its immediate recognition and international prestige.

> He was himself Hitler's greatest public enemy. . . . He was a brave man, but if he returned [to Germany], he would be killed. . . . Belgium suited him. He was more comfortable in small cozy countries (Holland was his favorite), but he wasn't safe from the Nazis. Unwillingly, he set off on his travels again, [and moved] to Princeton. . . .
>
> It was a kind of exile. There is no doubt that he, who had never recognized any place as home, sometimes longed for the sounds and smells of Europe. Nevertheless, it was in America that he reached his full wisdom and his full sadness.[14]

Before leaving Germany, Einstein was not only an outspoken critic of the Nazis, but he had begun to speak out against racism in America—the parallel to Nazi anti-Semitism and the Aryan-Superman theory was hard to miss.

In 1931, W.E.B. Du Bois, a founder of the NAACP and editor of its magazine, *The Crisis,* wrote to Einstein, still living in Berlin:

> Sir:
>
> I am taking the liberty of sending you herewith some copies of THE CRISIS magazine. THE CRISIS is published by American Negroes and in defense of the citizenship rights of 12 million people descended from the former slaves of this country. We have just reached our 21st birthday. I am writing to ask if in the midst of your busy life you could find time to write us a word about the evil of race prejudice in the world. A short statement from you of 500 to 1,000 words on this subject would help us greatly in our continuing fight for freedom.
>
> With regard to myself, you will find something about me in "Who's Who in America." I was formerly a student of Wagner and Schmoller in the University of Berlin.
>
> I should greatly appreciate word from you.
>
> Very sincerely yours,
> W.E.B. Du Bois[15]

Einstein replied on October 29, 1931:

> My Dear Sir!
> Please find enclosed a short contribution for your newspaper. Because of my excessive workload I could not send a longer explanation.
>
> With Distinguished respect,
> Albert Einstein[16]

The February 1932 issue of *The Crisis* featured the following article:

> *A Note from the Editor [Dr. Du Bois]:*
> *The author, Albert Einstein, is a Jew of German nationality. He was born in Wurttemburg in 1879 and educated in Switzerland. He has been Professor of Physics at Zurich and Prague and is at present director of the Kaiser-Wilhelm Physical Institute at Berlin. He is a member of the Royal Prussian Academy of Science and of the British Royal Society. He received the Nobel Prize in 1921 and the Copley Medal in 1925.*
>
> *Einstein is a genius in higher physics and ranks with Copernicus, Newton and Kepler. His famous theory of Relativity, advanced first in 1905, is revolutionizing our explanation of physical phenomenon and our conception of Motion, Time and Space.*
>
> *But Professor Einstein is not a mere mathematical mind. He is a living being, sympathetic with all human advance. He is a brilliant advocate of disarmament and world Peace and he hates race prejudice because as a Jew he knows what it is. At our request, he has sent this word to* THE CRISIS *with "Ausgezeichneter Hochachtung" ("Distinguished respect"):*
>
> TO AMERICAN NEGROES
>
> It seems to be a universal fact that minorities, especially when their individuals are recognizable because of physical differences, are treated by majorities among whom they live as an inferior class. The tragic part of such a fate, however,

lies not only in the automatically realized disadvantage suffered by these minorities in economic and social relations, but also in the fact that those who meet such treatment themselves for the most part acquiesce in the prejudiced estimate because of the suggestive influence of the majority, and come to regard people like themselves as inferior. This second and more important aspect of the evil can be met through closer union and conscious educational enlightenment among the minority, and so an emancipation of the soul of the minority can be attained.

The determined effort of the American Negroes in this direction deserves every recognition and assistance.

Albert Einstein[17]

(For the complete text of the Du Bois–Einstein correspondence, see Part II, Document A.)

Du Bois's request for a message from Einstein revealed that the African American scholar had a flair for public relations. Einstein's article brought *The Crisis* a rare, if small, headline in the *New York Times*: EINSTEIN HAILS NEGRO RACE.[18] Nearly twenty years later, another Einstein–Du Bois correspondence would bring even more momentous results, but in the fearful 1950s, there would be no press coverage.

On the eve of his move to America, Einstein joined the international campaign to save the "Scottsboro Boys," nine African American teenagers from Alabama, falsely accused of rape. Eight of them were sentenced to death in 1931. For Einstein, the Scottsboro defense was the first of several protests against racial injustice in the American legal system. For J. Edgar Hoover and his FBI, it was the first "Communist Front" listed in Einstein's file. But that is jumping ahead of our story.

CHAPTER 2

"Paradise"

Princeton seemed like another world to Einstein when he and Elsa arrived in October 1933. The small town's most striking first impression had to be the contrast with Berlin—the absence of gangs with swastika armbands roaming the streets, beating up Jews and Gypsies, attacking left-wing and trade-union meetings, and smashing shop windows. Princeton promised a safe haven—"a banishment to paradise," Einstein wrote in a letter to a friend. There was also a stuffiness to Princeton, emanating from the cluster of old, elite families and "society" people. Einstein would later describe it as "a quaint and ceremonious village populated by puny demigods on stilts."[1]

Nonetheless, he was clearly affected by Princeton's surface serenity: "Into this small university town, the chaotic voices of human strife barely penetrate. I am almost ashamed to be living in such peace while all the rest struggle and suffer."[2]

When people think of Princeton, they think of the university, which is world famous, Ivy League, rich, and conservative. One doesn't really think of the town. Actually, there are two Princetons: the central part is a borough, about 1.7 square miles with a population today of around fourteen thousand. The borough is surrounded by a larger township. Both Princetons are pretty and quiet and natural-looking and green, and each exudes a definite charm. Nassau Street, the borough's biggest and widest thoroughfare, separates a downtown commercial area from the university. The shaded campus, with its old trees and classic nineteenth-century buildings, stretches out in the distance.

It is easy to think of Princeton as some nondescript little college town. In some ways this may be, but if you're in the borough long enough you can begin to feel that America began in Princeton. George Washington slept there and it was the site of a battle considered by many to be a turning point in the Revolutionary War.

Princeton's early European settlers resided along a large stream. This stream was also a life source for the Leni Lenape, the Native American group that had long lived in the region. They called the stream Wopowog; the Europeans called it Stony Brook. The Lenape were nomadic and didn't believe in the concept of individual landowning. But European settlers quickly discovered the land was fertile and excellent for farming. Their settlement grew rapidly.

Princeton, originally Prince-Town, was either named after William, Prince of Orange, of the House of Nassau, or was so-named as a sister city of Kings-Town (today Kingston), which was not far away. Location also played a role in the town's development. Approximately the same distance from Philadelphia that it is from New York City, Princeton became a popular overnight stop for those traveling by stagecoach.

From its inception, the university played a fundamental role in the town's development. Like a magnet, it attracted people, money, and business. At its founding in 1746 in Elizabeth, New Jersey, it was called the College of New Jersey. The school moved to Newark in 1747, and finally to Princeton in 1756. The wealthy, powerful, and prominent individuals who ran the school often ran the town as well.

In the history of Princeton, the name Stockton is pervasive. The family estate, Morven, a museum now, is a permanent reminder of their presence. History made the name Richard Stockton an important, if confusing one—Richard became an extremely popular name in this family. Several were known well beyond the borders of their birthplace. One Richard Stockton, a Mississippi state judge, was killed in a duel at the age of thirty-one. A Richard Stockton Field was a U.S. senator from New Jersey. However, at least three Richard Stocktons made fundamental contributions to Princeton's, and in some ways the nation's, legacy. Historians consider the original Richard Stockton—for descriptive purposes, Richard Stockton I—to be the founding father of the town. He bought 6,400 acres of land from William Penn in 1696. The borough of Princeton is located almost in the middle of the land he purchased. He transformed a huge tract of this land into a thriving farm that exponentially added to the family's power and wealth.

His youngest son, John Stockton, initiated a family conversion from the Quaker to the Presbyterian faith in the 1750s. After personally donating forty wooded acres, he, along with three others, contributed a thousand pounds to help the College of New Jersey move to Princeton.[3]

John Stockton's eldest son, Richard Stockton II, a member of the first graduating class of what would become Princeton University, built the Stockton palatial estate called Morven, which in the 1950s became the official home of the governor of New Jersey; it was designated a museum in 1981 and opened to the public in October 2004. A successful lawyer and signer of the Declaration of Independence, Richard Stockton II was one of the tragic heroes of the Revolutionary War. Captured by the British, he died in 1781, never physically recovering from a brutal confinement. One of Richard Stockton's sons, Richard Stockton III, also known as "the Duke," represented New Jersey in the U.S. Congress. One of the congressman's sons, Commodore Robert Field Stockton, a naval officer, helped set up a government in California after the Mexican-American War and was active in the colonization society, a nineteenth-century effort to repatriate American blacks to Africa.* Later, the commodore also became a U.S. senator from New Jersey.

*At the beginning of the nineteenth century some began to feel that the most humane and/or expedient way to deal with the issue of racism in America was to send African Americans back to Africa. This belief spurred the birth of an organization that eventually became national called the Colonization Society. According to *The Journal of Negro History* 2, no. 3 (July 17, 1917), "The Formation of the American Colonization Society," 214, on November 6, 1816, Erkuries Beatty and Robert Finley led

In 1748 Jonathan Belcher, governor of the province of New Jersey, decided to make education a major focus of his administration to "better enlighten the minds and polish the manners of this and neighboring colonies."[4] After some dispute as to location, he settled on Princeton as the final location of the "College of New Jersey." Governor Belcher felt Princeton was "as near the center of the province as any and a fine situation."[5] Philadelphia architect Robert Smith was commissioned to create a building for the college. Smith built Nassau Hall, at the time "the second largest stone building in the colonies, it housed the entire college: classroom, dormitory, chapel, library, and refectory. Nassau Hall became a model for buildings at Harvard, Brown, Dartmouth, and Rutgers."[6] The college opened its campus in 1756. It changed its name to Princeton University in 1896.

The college's president, John Witherspoon, was both a great patriot and one of the individuals very much responsible for making the College of New Jersey one of the most prestigious schools in the nation. Witherspoon was born in Scotland. He graduated from the University of Edinburgh, one of the most respected institutions of learning in the world at the time. An ordained minister, he became a widely known leader of the evangelical or "Popular Party" in the Church of Scotland. Witherspoon's dynamic and extemporaneous speaking style mesmerized audiences. It was said in his time that only George Washington was more charismatic. Sought after and recruited for over a year, finally convinced by such notables as Benjamin Rush and Richard Stockton II, Witherspoon arrived in Princeton with much fanfare in 1767 to become the school's sixth president. Upon taking the position, he quickly realized that the college needed money; he traveled throughout the colonies but primarily in the South soliciting funds and recruiting students.

Colleges were somewhat different in the eighteenth century.

"the first Colonization meeting ever held in this country, which was in Princeton." See John F. Hagemen, *History of Princeton and Its Institutions* (Philadelphia: J. B. Lippincott, 1879), 223.

Much smaller than now, institutions of higher education mainly trained clergy. During Witherspoon's tenure the school employed only two other professors and three tutors. Witherspoon, in addition to managing the school's affairs and preaching twice on Sundays, carried a heavy teaching load and broadened and enriched the college's curriculum. He reportedly was the first person to use the word "campus" to describe college grounds. Under Witherspoon, as future U.S. President Woodrow Wilson later said, "Princeton became herself for a time . . . the academic center of the Revolution." An extraordinary number of Princeton graduates became notable statesmen: twenty-one senators, thirty-nine representatives, twelve governors, three Supreme Court justices, one vice president (Aaron Burr Jr.), and a president (James Madison), all within a period of about twenty-five years, from a college that seldom had more than a hundred students. Nine Princeton men were delegates at the Constitutional Convention, and five of these were Witherspoon's students.[7] Witherspoon himself became a signer of the Declaration of Independence.

While Witherspoon was president, a pivotal moment in the Revolutionary War occurred in Princeton. From 1775 almost to the end of 1776 the ragtag Continental Army, led by George Washington, was in continuous retreat. The British chased them from New York to the eastern part of Pennsylvania. For the colonial soldiers, rain, mud, hunger, cold, and a lack of adequate shoes and clothing wreaked more havoc than enemy bullets.

As the Continental Army crossed the Delaware into Pennsylvania, they were so weak and tattered that Washington wrote, "I think the game is pretty near up."[8] The British commander, Sir William Howe, decided to send his army into winter quarters. Howe was willing to wait until spring to kill any Continental soldiers who survived. Realizing this, Washington stopped running and planned an attack. Hessian soldiers, German mercenaries hired by Britain, manned an outpost in Trenton. Washington planned a surprise attack on the day after Christmas. At 8 a.m., as sleet fell, Washington's troops and artillery came from three different directions, totally surprising the Hessians. The Continental Army achieved victory in barely two hours of fighting that was fierce at times.

The emboldened Washington wanted to continue the offensive. In an impassioned speech he convinced his men to stay and fight, even though many of them were about to end their enlistments. On New Year's Day they recrossed the Delaware only to be cornered by British troops. It was getting dark. The British commander, Lord Cornwallis, held off his attack, telling his officers, "We'll bag the old fox in the morning."[9] During the night, Washington ordered a few men to stay behind to make campfires and noises as if they were busy building tents, while the bulk of his soldiers, using a little-known trail, headed toward Princeton, where two British regiments, cut off from the main force, were located. As day dawned Cornwallis realized the old fox had given them the slip. Soon he was informed that Washington had attacked in Princeton.

At one point, the story goes, Washington, thirty yards from enemy lines, ordered his men to fire. They did and the British responded in kind. Washington disappeared in the smoke. When the smoke cleared, Washington emerged unharmed but his crushed adversaries were running for their lives, taking shelter in and around Nassau Hall. On that memorable day, according to Woodrow Wilson, there was fighting in the streets and cannon fire aimed at Nassau Hall. Washington, who only a few days earlier had been beaten and in full retreat after crossing the Delaware, defeated the British and "changed the whole face of the war."[10]

A part of the story most people haven't heard is that black colonial Americans in the community of Princeton and beyond played a significant role in this critical victory. Several colonial African Americans, including many from the elite all-black First Rhode Island Regiment* fought in the battle. Black Revolution-

*The all-black First Rhode Island Regiment was composed of thirty-three freedmen and ninety-two slaves who were promised freedom if they served until the end of the war. They distinguished themselves in the

ary War veteran Oliver Cromwell recalled in the spring of 1852, at the age of one hundred, how Washington's army "knocked the British about lively."[11] Some of the fighting took place in Princeton's African American community. "Nineteen Hessian soldiers were killed on Witherspoon Street. For years after the battle residents spoke of being terrorized by a ghost of a Hessian soldier who was killed in the fight."[12]

In these times, many enslaved African Americans ran from bondage, spurred by principles of democracy, equality and liberty. A Princetonian who went by the name of Prime was one such individual. His owner, Absalom Bainbridge, supported the British. Bainbridge enlisted and served as a physician to the Loyalist troops stationed in Long Island. Prime, forced to travel with his owner, bolted and returned to Princeton. There he was advised to join the Continental Army, which he did with the hope that this service would earn him freedom.

However, after the war, Prime was not legally released. He lived a quasi-free existence until 1784, when he was captured by slave hunters. At that time his legal representatives filed a petition for manumission. Prime's lawyer argued that Absalom Bainbridge had lost all property rights when he became an enemy of the state and joined the British Army. The court ruled in Prime's favor, and he was finally freed. What happened to him afterward is a mystery; to date no records have surfaced that document his life. Princeton slave owner Absalom Bainbridge, on the other hand, rented a house in 1774 from Job Stockton, a brother of Richard II. Though it was confiscated when he joined the English in the war, today his landmark house still bears the Bainbridge name. Named after Absalom's son, William Bainbridge, a hero in the War of 1812, it is now the home of the Princeton Historical Society.

"No college has turned out better scholars or more estimable characters," George Washington wrote after the war of New Jersey's illustrious school in Princeton. He was a lifetime sup-

---

Battle of Newport. However, the regiment was all but wiped out in a British attack at Yorktown.

porter, even though his stepson was expelled from the school. Another U.S. president, James Madison, lent his support to the school by coming to the campus and helping to raise funds.

In the second half of the nineteenth century, two of Princeton University's presidents made important contributions to the school and the community. John Maclean Jr., the only Princeton president who was a bachelor, played an instrumental role in the development of the town's public school system; and in the 1880s James McCosh, a Scottish-born Princeton president, focused on constructing impressive buildings that has made the campus one of the most beautiful in the country.

In 1902 a former student who was now the school's most popular and highest-paid faculty member was elected unanimously to be Princeton's president. Though a Presbyterian and son of a minister, Woodrow Wilson was the first Princeton president who was not a clergyman. He is credited with modernizing the school. Many of the reforms he instituted have become standard in higher education. "In his first report to the trustees Wilson proposed a $12.5 million program to transform Princeton into a full-scale university. At the time this was a staggering sum but the trustees approved it immediately. Wilson created administrative departments of instruction with the heads reporting to him. He instituted the now common system of core requirements followed by two years of concentration in a selected area for students."[13]

Wilson left Princeton in 1910 to run for, and win, the governorship of New Jersey. Two years later, he was elected president of the United States—the first and still only U.S. president to have a Ph.D. Wilson was also the first southern president since the Civil War.

When Albert Einstein first visited the university in 1921 to lecture and receive an honorary degree, John G. Hibben, Woodrow Wilson's successor, was the university's president. He has often been described as more of a coordinator and mediator than a dynamic leader. By 1933, the year Einstein moved to Princeton, Harold W. Dodds was just beginning his Princeton presidency.

Dodds's long tenure would last until 1957, two years after Einstein's death.

Einstein may have been aware of the quantity and quality of history and the historical figures who preceded his arrival. He may also have been surprised when he discovered another aspect of Princeton life.

We can picture Einstein as he explores the streets of his new hometown, walking from the university to Nassau Street with its stores and student eateries, and making his way down Witherspoon Street. Perhaps Einstein first noticed the scarcity of well-paved sidewalks here, or perhaps the sudden presence of black people and the disappearance of whites. In either case, it doesn't take him long to learn that black Princetonians live in a separate community, send their children to a separate school, sit in separate sections in the movie theaters, and cannot enter most stores on Nassau Street.

This is the other history of Princeton, one that is not usually cited in American history textbooks. Benjamin Rush, a signer of the Declaration of Independence, the preeminent physician of his day, and a Princeton graduate (class of 1760), had this remembrance: "In the days of Ezekiel Foreman and Alexander Macdonald, Prince-town's main street was lined with shops and taverns and the village had its own post office and an active middle class of physicians, lawyers, and tradesmen. There were regular market days with sales of slaves, cattle, horses, and sheep. A daily stagecoach transported travelers between New York and Philadelphia stopping in Prince-Town."[14] "Sales of slaves" may have ended, but the plantation mentality continued to dominate Princeton's elite—both in the town and the university—not just for a few years or decades, but for another one hundred years. This is the other history of Princeton—and it's not so pretty.

CHAPTER 3

The Other Princeton

> A writer to *Harpers Monthly* in the 1850s said that if in passing through New Jersey a traveler came to a place where there were two darkies to every white man and two dogs to every negro, he might be sure he was in Princeton.
>
> —Lloyd Brown, *The Young Paul Robeson,* 22

They say Witherspoon Street is named after the path that John Witherspoon took from his home to the university and back. In modern times and for over a century, a walk north on this block, once an Indian trail, leads to the African American community. Before it was Witherspoon Street it was Hill Road, also known as Rocky Hill, African Lane, or African Alley. Witherspoon and the neighboring Clay, Maclean, Quarry, and John Streets look today like other parts of the town, sleepy and tree-filled. Some of the houses may be a little smaller, not as modern or pristine, but there's nothing to indicate that this little section has a history more like that of Selma, Alabama, than New Jersey. "The black community was separated and set apart," Princeton Public Librarian Terri Nelson said, commenting on the historic social conditions in town. "At the same time, they were completely involved in the very fabric of Princeton's daily life." Paul Robeson, who was born in Princeton, called it "the northernmost town in the South."

There are black people today in Princeton who remember not being allowed in certain restaurants and not being served in certain stores. At the movies they were required to sit in the "colored" section, and they had to go out of town to attend high school. Princeton was different from nearby towns like New

Brunswick or Trenton, which surely had racism but not such a long history of ingrained Jim Crow-ism. "In New Brunswick," as one resident put it in the 1940s, "you went to school based on your address, not your skin color."[1]

There have been tales that Princeton University was built with slave quarters on campus. Most historians believe there is no evidence of that. However, there is no doubt that the university, a bastion of intellectual thought, was at the vanguard of instituting and perpetuating harsh racial segregation in the town.

Bruce Wright, a writer, activist, and former judge in New York's Criminal Court, who had developed a reputation during his sixteen-year tenure for his progressive and controversial rulings, was born in Princeton. In 1936 he received a full scholarship to the university but was denied admission because he was African American.

"I had thought that I would attend Princeton University," Wright wrote.

> A scholarship arranged for me by one of my high school instructors had brought joy to a family that had no money. We were, after all, in the midst of the Great Depression.
>
> I stood in the registration line full of hope. The sun was shining and the green lawns of the University were beautiful to behold. In my innocence I was untroubled when an upperclassman, an orange arm-band identifying his status, asked me to follow him to the office of the director of admissions.
>
> I was ushered into the presence of Radcliffe Heermance. This man of Falstaffian girth towered over me. Light shafted through the leaded windows of the office. Heermance stood there, as though surrounded by a divine radiance. He was the first man to address me as "Mr. Wright." His next words, however, would destroy much of the child and educate me beyond anything I would ever learn in a classroom. "We did not know you were colored when the scholarship was arranged."[2]

From the time of John Witherspoon, the college had made a concerted effort to cultivate a relationship with the southern elite. Perhaps because Princeton was farther south than Harvard, Yale, Columbia, or Dartmouth, this southern strategy worked. A study of the decade of the 1840s showed that there was only one year when the percentage of southern students at Princeton was below 40 percent. In 1848 it was as high as 51 percent.[3]

During the Civil War, thirty-five students died fighting for the Union and thirty-five for the Confederacy. Though several southern Princeton students left the school in those heated, tension-filled days, the breakup was temporary. Princeton's southern tradition blossomed again soon after the war was over. As a member of the class of 1928 put it, "Princeton is popular through the South because it is the one eastern school which does not enroll Negroes."[4]

The number of African Americans who attended classes at Princeton from 1756 to 1947 and received Princeton undergraduate degrees amounts to precisely one. That would be John Leroy Howard in 1947.* In 1948 when the New Jersey legislature passed a law that forced the end of segregation in Princeton's public schools, the university, under the leadership of President Howard W. Dodds, took a stand against the legislation. Not surprisingly, Princeton was the last Ivy League school to integrate.

Racism in Princeton and the victimizing effect it had on black peoples' lives did not begin with the university. Enslaved Africans, purchased from Dutch slave traders, had been part of the early Stony Brook settlement. When Richard Stockton I arrived

*Nearly three-quarters of a century ago a well-known Trenton minister, the Reverend Irwin W. L. Roundtree, who for many years was associated with the A.M.E. Zion church, became the first Negro to earn a Princeton University degree. In 1895, the year before the College of New Jersey officially became Princeton University, Roundtree, a graduate of Lincoln University and Princeton Theological Seminary, received a Master of Arts degree from Princeton. (From the Department of Public Information, Princeton University, July 18, 1969.)

in Princeton, he owned at least seven enslaved Africans. Years later, enslaved laborers built Morven, the Stockton family mansion. They also constructed brick slave quarters for themselves on the estate.

John Witherspoon, a slave owner himself, reportedly had at least six African-descended students in his charge while he was Princeton's president. Two of his colonial African students, John Quamine and Bristol Yamma, were being trained to lead groups of African Americans back to Africa, forerunning the colonization movement that began in Princeton and became widely popular in the 1820s. The advent of the Revolutionary War disrupted the plan, as well as Quamine's and Yamma's continued education.

The best-known African American student in the Witherspoon era, John Chavis, began his Princeton education in September 1792. It was the Reverend John B. Smith who recommended that Chavis receive a scholarship called the Leslie Fund, which had been set up that year for the education of poor students who would agree to work in the Presbyterian Church. Although Chavis did not graduate (he is now on the official roll of nongraduates), he fulfilled the terms of the Leslie Fund and became a licensed Presbyterian minister, preaching extensively in North Carolina and Virginia. Chavis was quite active until 1832 when North Carolina, alarmed by the Nat Turner slave revolt the previous year, forbade Negroes to speak in public under any circumstances. In earlier years, when Chavis had a school in Raleigh, he is believed to have had both black and white students, unusual for an African American teacher in North Carolina at the time.

In Princeton in the early decades of the nineteenth century, eighty or ninety blacks regularly attended services at the First Presbyterian Church on Nassau Street. They were allowed to sit only in the balcony. When fire gutted the church and the building was restored, a controversy ensued. A letter written by Dr. James Alexander, a Princeton professor and noted member of the church, explained the congregation's feeling.

> We have a new and handsome edifice. While it was building the negroes worshipped apart, in a little place of their own. The majority of the pewholders wish them to remain a separate congregation. . . . If they come back, they will take up about half the gallery. I am clear that in a church of Jesus Christ there is neither black nor white; and that we have no right to consider the accident of colour in any degree. Yet I think the blacks very unwise in insisting on such a privilege now. Some years ago there would not have been the slightest difficulty in admitting them, but in consequence of the abolition movements the prejudice of the lower classes of whites against the blacks has become exorbitant and inhuman.[5]

The black congregants were given the option to form their own church or be unwelcome in the newly rebuilt one. After intense negotiations, they opted for their own, largely funded by a loan from their former white congregation. They eventually named their new church Witherspoon Presbyterian.

In the run-up to the Civil War, Princeton, like the rest of America, was divided. Though many hated African Americans and the war, there were those who supported the abolitionist cause. General David Hunter was one such individual. Also known as Black Dave, he became Abraham Lincoln's friend and an abolitionist general. In the spring of 1862 in South Carolina, Hunter was accused of raising and equipping a regiment of African American runaway slaves. When the House of Representatives asked for an accounting, he responded: "No regiment of fugitive slaves has been or is organized in this department. There is, however, a fine regiment of persons whose late masters are fugitive rebels."[6] At the time Hunter was assembling this force, the North was still pretending the Civil War had nothing to do with slavery. Up to this point few blacks, if any, wore Union uniforms. When asked about the legality of his actions, Hunter replied: "I conclude that I have been authorized to enlist 'fugitive slaves' as soldiers, could any be found. No such characters have, however, yet appeared within our most advanced pickets, the loyal slaves everywhere remaining on their plantations welcome us, and supply us with food, labor, and information. It is the masters who have, in every instance, been the 'fugitives'—rifles in hand, dodging behind trees, in the extreme distance."[7]

On May 9, 1862, the general issued a proclamation that ended slavery in Georgia, Florida, and South Carolina. President Lincoln assured the nervous Congress that Hunter's proclamation had no official authorization. A longtime abolitionist, General Hunter was also a Stockton. His mother, Mary, was the daughter of Richard Stockton II, the signer of the Declaration of Independence.

Because New Jersey did not form black or, as they called them, Colored Troop regiments, black Princetonians were forced to enlist in other states. Despite the inconvenience, after Lincoln's Proclamation of Emancipation in 1863, twenty-two black men from town joined the Union navy and more than sixty-five the army.

The end of the Civil War and the ensuing years did not bring about a rise in status for black Princetonians. They were still largely unaccepted and unwelcome throughout town. As a result, in the Witherspoon Street community you could find black-owned florists, beauty parlors, and clothing and candy stores. There was Griggs Restaurant, Gray's Restaurant, and De Paur's Restaurant. Black Princetonians printed their own newspaper, *The Citizen*, created the Colored YMCA, Elks Club, and even their own schools, notably the Witherspoon School.

In 1858, the Witherspoon School for Colored Children, already in operation for a decade, opened in a larger building located on the corner of Maclean and Witherspoon Streets. With its strong focus on academics, the school became an icon in the community, one where African American teachers would educate young black Princetonians for more than a century.

The school's legacy, like much else in Princeton, was directly connected to the town's most prominent family, the Stocktons. Betsey Stockton is the person most associated with the founding, development, and success of the Witherspoon School. This dynamic woman, and perhaps the most unusual of the Stocktons, was an African American member of the family. Born in 1798 in an era when most women were confined to family and home, Betsey Stockton was a traveler, an intellectual, a slave, a seamstress, a nurse, a writer, a teacher, and a leader in her community.

Ashbel Green, who raised Betsey Stockton, called her "wild and thoughtless, if not vicious," until about the age of thirteen.[8] Her childhood seemed extremely difficult. Betsey Stockton may have been simmering with anger. Though Robert Stockton was her father, their relationship was shrouded in secrecy, because Robert Stockton's wife was not Betsey's mother. Her mother, Ceila, was Robert Stockton's slave. Ceila, soon after her daughter's birth, ran away from her child and from slavery. Betsey Stockton, at the age of six, possibly as "part of a settlement of a civil case against Robert Stockton's estate,"[9] ended up as property of Ashbel Green, who was the eighth president of Princeton University,* and Elizabeth Stockton Green, Robert Stockton's daughter. Ashbel and Elizabeth Green had no desire to keep Elizabeth's half-sister enslaved, and Betsey became a "member" of the household.

Educator Ashbel Green recognized that Betsey was quite intelligent and had the ability and desire to learn. By giving her access to his vast library and hiring tutors, Green facilitated her education. At the age of twenty-five, Betsey traveled to Hawaii as a missionary. Spending several years there, she easily learned the local language, founded a school, and discovered that her mission in life was teaching.

After her experience in Hawaii, she spent a few years as an educator in Philadelphia before returning to Princeton. When the Witherspoon Presbyterian Church opened in 1837, Betsey Stockton's name appeared first on the list of ninety-two commu-

*Ashbel Green was president of Princeton from 1812 to 1822. His autocratic and old-fashioned approach to his Princeton presidency incited several major student outbursts. During one, a gunpowder charge was set off in the main entrance of Nassau Hall, cracking the walls from top to bottom. After a dispute with the trustees about the tenure of his son, who was a faculty member, Green resigned.

nicants, an obvious reflection that she had some standing in the congregation. Betsey Stockton's major mission, however, was to establish an African American school in the borough. She began teaching in Princeton in 1837 and her small one-room school eventually grew into the Witherspoon School. The force of Betsey Stockton's spirit and legacy in Princeton continued long after her death in 1865.

Despite academic achievement by blacks, for the most part white Princetonians remained as racist as ever. Many pined for the old South and fervently believed that African Americans were incapable of coping with freedom. An article published in the university newspaper *The Princetonian* in 1895 entitled "One Result of the New System" reflects this view.

> Saunter down Witherspoon Street with me, friend, past the rows of tumble-down shanties and more tumble-down inmates. Let us walk on a short distance towards Rocky Hill and pause as we come to an abrupt bend in the road. The conglomeration of shingles, warped boards, mud and stones, which lies before us is too small for a barn and too large for a hencoop. It may be a house? Yes, there is the chimney, or rather what is left of it. Abraham (who lives there) is in the laundry business, and believes in the "Division of labor." That is, he divides the washing between his good wife Hannah and his elder daughters, and divides his own time between the college offices, where he collects his drawbacks on washing, and the saloon.[10]

With its growing student body, campus, kitchens, and off-campus eating clubs, Princeton University increasingly became a source of income for many of Princeton's African Americans. The jobs were mainly labor-intensive and menial. "Many of the men worked in those eating clubs on the Avenue, Prospect Avenue," said Estelle Johnson, a longtime resident. "They got to know some of those fellows, but that was as far as it went. Those students could talk to them and kid around with them but the man still had to call the student Mr. So and So."[11]

A host of black Princetonians established entrepreneurial relationships with the university. There was William Sport Moore. His grandson Donald Moore recalls that "the students at Princeton were not given a lot of money, but they were given unlimited credit accounts at clothing stores. They would want to go to New York, so they would take their clothes to my grandfather, who would buy them. They'd take the money and go to New York and have a great time, and my grandfather would sell the clothes to whomever. And a lot of this stuff was brand new."[12] In addition to selling clothing, Sport Moore bought and sold student furniture. Kid Green, reportedly the first black man to own an automobile in Princeton, made money taking care of polo ponies for students.

African American vendors had long been an essential part of Princeton campus life. Peter Scudder, perhaps the best known of these vendors, worked as a bootblack and sold apples and ice cream to college students in the 1840s. He accumulated some wealth and eventually became a property owner, an oddity then for an African American in Princeton. Scudder, commonly known as "Peter Polite," owned a popular ice cream and confectionary shop on Nassau Street.[13] Almost twenty years later, Jimmy Johnson created quite a stir when he arrived in Princeton as a runaway slave. Southern bounty hunters attempted to return him to captivity but several Princeton residents interceded in his behalf and purchased his freedom. Johnson's wit and electric personality, as much as anything, saved him from a chattel existence. As an older man, in the 1890s, he was a campus fixture, selling candy and peanuts from a large wheelbarrow and entertaining customers and passersby with his humorous tales. When Johnson died in 1902, Princeton students raised money to get a tombstone for his grave. Perhaps the last of the line of Princeton's memorable African American vendors was William Taylor, somewhat derisively nicknamed "Jigger." Taylor and his wagon were familiar sights as he sold his famous bacon buns, hot dogs, apples, and candy on Nassau Street and at all of Princeton's sporting events well into the 1940s.

For the first two hundred years of the school's existence,

Alexander Dumas Watkins was the closest thing to an African American teacher that Princeton University ever had. "He was the only Negro who ever instructed Princeton undergraduates in their regular curriculum work."[14] He did so for about eight years. An expert in histology and biology he, as the *New York Times* reported, "was allowed to tutor some of the less intelligent students."[15]

Watkins's sudden death in 1903 came on the eve of a new wave of repression of African Americans in Princeton. The beginning of the twentieth century brought a decidedly more racist climate throughout the nation, with parades by thousands of Ku Klux Klan members in the South and in other states, including New Jersey, where on Labor Day 1924, ten thousand Klansmen paraded through Yardville and Mercerville and then on to Hamilton where they held a public "Klanvocation" rally.* Even in Washington, D.C., the KKK was "sufficiently powerful to parade unmasked in the shadow of the Capitol." Between 1900 and 1914 more than 1,100 African Americans were lynched in southern states.[16] In Princeton, the already racist atmosphere was exacerbated by Woodrow Wilson. Whatever else he may be known for in history books, Wilson was, to put it simply, a racist.

Born in Virginia in 1856, Wilson demonstrated a nostalgic longing for some of the worst aspects of the antebellum South. When the movie *The Birth of a Nation* came out in 1915, thousands of people around the country were picketing because of the film's glorification and justification of the KKK. Woodrow Wilson, who was then president of the United States, arranged a special White House showing of the film. After seeing it, Wilson said: "It's like writing history in lightning. My only regret is that it is all so terribly true."[17]

Wilson would buy peanuts from a black man or let him shine his shoes, but under his leadership Princeton's world of academia was no place for African Americans, and he wasted no time making this perfectly clear. A Woodrow Wilson story, widely

*The Klan claimed more than three million members by the 1920s. See *Life* magazine's *Century of Change* (Boston: Little, Brown, 2000), 216.

believed in the black community, may not be true. Alice Satterfield, a longtime member of Witherspoon Presbyterian Church, recounts it this way:

> Paul Robeson's father [William D. Robeson] was minister of our church for twenty-one years. That was a time of deep Jim Crow-ism and his father was always fighting for the rights of Negroes. His son Ben [Paul's older brother] was accepted at Princeton, but when they found out he was black they rejected him. There's a story that Paul's father had a confrontation with Woodrow Wilson who was then president of the University. Wilson talked to the Presbytery and Robeson was dismissed from his job and position as a minister.

But Lloyd Brown, in his book *Young Paul Robeson,* indicates that it is more likely that William D. Robeson lost his ministry because of the Wilmington [North Carolina] Riot of 1898. Angered by years of black empowerment and interracial government, the former slaveholding community took steps to reassert its hegemony. "The high point in the intimidation came in Wilmington where organized terror, personally participated in by 'the best people' deposed, by murder, the city administration on the day following the election. After months of planning, buying guns, organizing, whipping up race hatred, the assault came."[18]

Officially, thirty African Americans were killed. Many believed that the real number was closer to three hundred. The complete lack of response by local and federal authorities angered and galvanized African Americans throughout the nation. On November 29, a large group of black Princetonians assembled at Witherspoon Presbyterian to meet about racial violence. Reverend Robeson was one of the main speakers. He and two others were chosen to be delegates at a followup event in Philadelphia. Apparently, this assemblage and call to action (even though they advocated nonviolence) offended Princeton's ruling establishment. The black Presbyterian Church was controlled by an all-white governing body of local Presbyterians. On November 8, 1900, the Presbytery, as they were called, voted to oust Robeson from his church, despite his popularity with his congregation.

Paul Robeson wrote the following recollection of his father, William D., in his book *Here I Stand*: "Though a man of ordinary height, he was very broad of shoulder and his physical bearing reflected the rock-like strength and dignity of his character. He had the greatest speaking voice I have ever heard. It was a deep, sonorous basso, richly melodic and refined, vibrant with the love and compassion, which filled him."[19] After his dismissal, William Robeson endured several years of poverty. He resurfaced in Westfield, New Jersey, and became minister of the African Methodist Episcopal Church. At the age of sixty-two, he transferred to a larger church, Saint Thomas A.M.E. Zion in Somerville, New Jersey, where he ministered until he died at the age of seventy-three in 1918.

For most of the twentieth century—from Woodrow Wilson's presidency of Princeton beginning in 1902 until the civil rights movement of the 1960s—the prestigious university that dominated the town was its first fortress of segregation and racism. (Einstein's research home, the Institute for Advanced Study, was independent and apart from the university.) More than other Ivy League colleges, Princeton continued to attract a high percentage of students from southern states. The *Princeton Herald* "explained" that admitting black students to the university, while morally justified, would simply be too offensive to the large number of Princeton's southern students.[20]

After Bruce Wright was denied admittance to the university in 1936, when the school's officials discovered he was black, Wright wandered through Princeton "from one block to another." Describing an incident he felt represented white Princeton's racism, Wright says, "I noticed a white man sitting with his chair tilted back against the wall of a bar, listening intently to a radio playing within. Suddenly the man leaped to his feet, virtually in front of me. He slammed one fist violently into his other hand

and yelled, 'At last, they got that black son-of-a-bitch!' I later learned that Max Schmeling had knocked out Joe Louis in their first fight."*

Wealthy, white, and southern, Princeton students created more-than-usual tension, grief, and terror for the town's African American population. At one point early in the twentieth century, the black community even set up a curfew to protect its children. "I remember what those students used to do for the bonfire, the victory bonfire [when a Princeton athletic team won a game]," Emma Epps, one the principal and long-term leaders of Princeton's black community, told an interviewer in 1977. "They'd come down in our neighborhood and tear off our porch to burn up for their bonfire. They'd come down here and steal everything for that fire.""Oh, they were terrible," she continued, "and we had a curfew just for that reason. You'd see us running. Mr. Kilfoyle'd go, 'Run, Emmie, run!' And that curfew bell would ring at 9 p.m., [especially] in the spring; it was still kind of light at 9. They were terrors."

A similar picture emerges from a 1950 interview with Paul Robeson, a contemporary childhood neighbor of Emma Epps, describing his recollections of what black people faced in Princeton: "Almost every Negro in Princeton lived off the college and accepted the status that went with it. We lived for all intents and purposes on a Southern plantation. And with no more dignity than that suggests—all the bowing and scraping to the drunken rich." His words, "the drunken rich," strikingly echo the curfew-terror story. The ringing of the bells at 9 p.m.—probably from the church—meant not only that black children had to be at home, it meant a mob of drunken white racists was descending on the neighborhood.

"Oh, how I remember those bells!" said Emma Epps.[21]

*Wright, *Black Robes, White Justice,* 37. Two years later, in 1938, the two boxers fought a second match in Yankee Stadium, with the whole world watching and Hitler openly rooting for the German heavyweight to demonstrate yet again the superiority of the Master Race. Louis knocked Schmeling out in the first round.

Somewhat to his embarrassment,. Einstein became in the United States, as he had been in most of Europe outside Nazi Germany, a living metaphor for *genius,* "the smartest man on the planet." He was the science superstar and the most famous man in Princeton. Yet Einstein only had to walk a few blocks down Witherspoon Street to be reminded of how fleeting freedom could be. "The worst disease" in American society, he wrote, is "the treatment of the Negro. Everyone who is not used from childhood to this injustice suffers from the observation. Everyone who freshly learns of this state of affairs at a maturer age feels not only the injustice, but the scorn of the principle of the Fathers who founded the United States that 'all men are created equal.'"[22]

CHAPTER 4

Witherspoon Street

When they came to Princeton to make the movie *IQ* [a 1994 movie starring Walter Matthau as Einstein], I wanted to be in it. I remember dressing up all preppie and going for an audition. They ask you all these questions. I told them I used to *take walks with Einstein,* thinking that would get me in.

They had one black person in the movie—a waitress.

—Shirley Satterfield, a longtime African American resident of Princeton

In the fall of 1946, people in African American communities caught up in the Levi Jackson craze flocked to cheer him on. Florence Taylor, then a Brooklynite, believes this may have been the year she and her husband went to Yale University to see a football game. Florence knew nothing about football but remembers a packed house and the crowd going crazy chanting "Leeeeeviiiii, Leeeeeviiiii," as the electrifying African American running back, the star of Yale's team, busted color barriers and stereotypes. After the game she and her husband drove down to Princeton to drop a friend off at home.

Florence remembers that at one point in her friend's house there was a flurry of excitement. Everyone ran to the window. It was an Einstein sighting. Albert Einstein, the preeminent Princetonian, frequently used to walk down the block, which was in the heart of the black community. "He looked like a very ordinary type person," Florence recalled, "dressed casually—his hair was kind of all over the place."

Although it has never been reported by his legion of biographers, Einstein spent a good deal of his walking-talking time in Princeton's African American community. "He felt comfortable here," black Princetonian Henry Pannell explains. "We treated him like an ordinary person."

More than half a century later, memories of Einstein still resonate among residents of the community. Whether it was sitting on a neighbor's porch or a neighbor's steps, giving candy to a group of kids, a conversation here, a friendship there, a laugh—Einstein made himself part of their journey down Witherspoon Street. Below, a number of black Princetonians share their memories. (See Interviews section in the Bibliography for a list of the participants.)

Callie Carraway Sinkler: When we were girls, me and my sister Lili [Taylor] used to watch Einstein walking up Witherspoon Street, the main street in Princeton's black community. See, he cut across from his home on Mercer Street, so he would walk up on Witherspoon going to the hospital.

Morris Boyd: Yes, I saw Einstein often walking on Mercer Street . . . sweater hanging down, sandals. . . .We heard that Einstein's stepdaughter wanted to give him a car, but he preferred walking.

Penney Edwards-Carter: I remember Einstein riding his bicycle and also walking down the street in the early 1950s, giving us kids candy.

Fanny Floyd: Well, I used to see Einstein on Nassau Street. Usually he was walking on the side of the cemetery.

Lloyd Banks: Einstein was unusual for a white person to a degree—he wasn't bothered being in the black community. We would run out—as children we always ran out and talked to him and he would stop and talk to us. We would be shouting—"Dr. Einstein, Dr. Einstein"—and he would stop and take a few minutes with us. He took the time to talk to us. He was very friendly. It seems as though it was almost every day, or at least every week. He always walked.

Rod Pannell: He used to stand on the corner of Jackson and John Streets. . . . We lived just two doors in on Jackson. . . . I remember his khakis and sweatshirt, and sometimes he'd stand on that corner for quite a while, pondering the universe. He might have been thinking about the quantum theory or something like that. . . . He was almost like a part of the community.

Henry Pannell: I agree. It was a place he could feel really free.

Wallace Holland: I went to Einstein's house one day. . . . I was thinking about what happened to the smoke from a fire or a cigarette when it goes up in the air and goes beyond us. Where did the smoke disappear to? I wondered about it and decided to go ask Albert Einstein. He probably could tell me. . . . I didn't know anybody else who would be better to ask. I was ten or eleven years old. I went to his house—everyone knew where he lived—and his sister* opened the door. I said I want to speak to Mr. Einstein. He wasn't at home but I had no problem going by to see him. He was always talking with kids when he walked through the neighborhood.

*Maja Einstein arrived in Princeton in 1939. She had been living with her husband in Italy until Mussolini gave in to pressure from Nazi Germany and adopted anti-Semitic policies in that country. In 1938, the wife and children of Einstein's cousin Robert were brutally murdered by Italian fascists while Robert was away from home. When he returned and found what had happened to his family, he took his own life. Einstein immediately wrote to Maja in Florence and suggested that she come to "visit" him in America. She remained at the Mercer Street house until her death in 1951.

Timmy Hinds: Well, I remember seeing Einstein when I was twelve or thirteen. It must have been around 1952. He would walk down John Street. He'd walk down to Barclay's Ice and Coal Company where he'd get an armful of wood and then walk back to his house.

Shirley Satterfield: Or he'd be walking down John Street—he wouldn't be afraid to walk in the black community.

Eric Craig: There was just no racism about him. He came from that kind of situation, with what happened in Germany. I remember thinking that he must have experienced a terrible kind of race hatred against the Jews in Germany that made him more race conscious than most white folks in Princeton. We were actually thrilled when we saw him because our parents had told us this was somebody important. He'd walk down Witherspoon Street mostly every day—as children, we would run out—our parents told us he was famous. He'd always spend time—that was back in 1942, '43, and '44. I remember his long hair.

Evelyn Turner: He would walk slowly, with his hands behind his back. In those days, there were no buses going through town, just buses from Trenton that went down Nassau Street.

Mercedes Woods: I was a little kid about nine. I used to talk to him up on Nassau Street, up on the common square. He'd be sitting on the bench. He was a very nice man. He would have his hair uncombed, no stockings on, his shoes unlaced. He was a very funny man. I talked to him often. Einstein was very friendly.

Albert Einstein lived in a house at 112 Mercer Street in Princeton. For most of that time he shared it with his sister, Maja, his stepdaughter, Margot, and his secretary, Helen Dukas. One of the domestic workers they hired was Lillie Trotman, who still lives in Princeton. Mrs. Trotman suffered a stroke in 2002 that has left her largely unable to talk, but her daughter, Mary Trotman, and granddaughter, Lena S. Sawyer, shared their recollections:

Mary Trotman: My mother told me that Einstein helped pay for an African American young man from Princeton to go to college.

Another story Lillie would tell is that Einstein would sometimes come home and find the front door locked and—forgetting that the back door was open—would call the police.

Lena S. Sawyer: My grandmother worked as a domestic for Einstein. She knew details of his home life that were not part of what I'd been taught in school.

When Professor Einstein had visitors, they sat and ate in the dining room; she listened from the kitchen.

I remember that she said, "He was a friend of black people" and "He was a good man." Other details I remember were that he did not wear socks and that his wife was sick. I had a hard time understanding how she could characterize him as a good man and a friend to black people, since I remember I asked her if they ate together. For me, at age twenty-two, this was a sign of someone being on equal relations and thus "good." I had a hard time reconciling that he could be good and still have her sit in the kitchen and eat. Anyway, I remember she laughed and said of course she ate in the kitchen when he had people over. She was, after all, his help. I do not remember her saying why exactly she characterized him as a good man, but she said that he was a Jew and had experienced hardships in Europe and could understand black people's situation. But how this was expressed, I do not know. For example, I wonder—did he give her better wages? Treat her with respect and dignity? Did he give her material things?

I can remember in college my grandmother coming to me with clothes from one of her employers—Ms. Overton! For my grandmother, this was a sign of kindness. However, when in my second year of grad school she wanted me to finally meet Ms. Overton and "thank her" for these clothes, I had a hard time swallowing my anger at this woman—fit and trim from her tennis and leisurely life. She called my grandmother Lillie and my grandmother called her "Ms. Overton." It was also clear in our meeting that my grandmother was in a lower position of power than Ms. Overton. So I have a hard time understanding my grandmother's judgment of Einstein—what did she call him, I wonder? What did he call her?

Along with their memories of Einstein walking through the African American community, some of Princeton's black residents remember how they or their family members spent time with the scientist, chatting or strolling together.

Paul Hinds (from "Paul Hinds Leaves a Changing Field," July 16, 1985, *Packet Publication*): "I have lived among the eggheads. I sat down and talked to Dr. Einstein when I used to pick up garbage from him. I didn't just pick up the garbage and leave. I had to talk to the people—sit down and have a cup of coffee."*

Albert Hinds: My brother Paul had a garbage company and he picked up Einstein's trash. Einstein would invite him in often to talk.

Timmy Hinds: My father [Paul Hinds] would talk with Einstein's sister Maja when he went by to pick up the trash. He

*"I speak to everyone in the same way, whether he is a garbage man or the president of a university": Einstein to Peter A. Bucky. (See Part II, Document J.)

worked at Bamman's grocery store most of his working life before he owned and operated the small garbage disposal company and picked up Einstein's garbage (among others').

And I remember my father was a nature lover. He liked to read John McPhee and he loved nature and flowers. When he was out collecting trash, he'd pick one flower from a lawn and transplant it to a plot in our yard. Then the next day, he'd do it again with another flower, and pretty soon he had his own flowerbed. He also played football in high school.

Albert Hinds: My sister Violet Hinds Jones frequently worked for a family who were neighbors of the Einsteins.

My sister and Einstein would hold hands and walk up Mercer Street daily. I learned that from my niece.

Shirl Gadson: My mother [Violet Hinds Jones] was a member of the Friendship Club.* Her father was from British Guiana and worked at the university as a valet for the class of 1923. His picture is in that yearbook. Violet's mother was a Princetonian. They had five girls.

My mother didn't tell me much about Einstein. She just said, "I knew him." Once, talking about him, she said he was "a great friend." She did tell me they used to meet each other after she got off from work. He was standing there waiting at a certain spot. Then he'd say, "We seem to meet at the same place every day." Sometimes he would say, jokingly, "We've got to stop meeting like this."

*The Friendship Club, founded in 1932 by six leading women in Princeton's African American community, helped students (with scholarships), promoted arts and artists (sponsoring concerts by artists including Dorothy Maynard, Duke Ellington, Paul Robeson, and others), and provided aid for a variety of causes such as migrant farm workers in nearby communities. Founding members were Emma Ashe Banks, Bertha Hill Brandon, Emma Greene Epps, Carrie Dans Pannell, Lucy Ruffin, and Louise Skipworth. It was affiliated with the N.J. State Federation of Colored Women's Clubs and the NAACP.

"Yes, we do."

"Well, we can walk together."

"What did you talk about?" I asked her.

"We talked about the students, about the town. Well, maybe not every day but frequently. Sometimes we only walked a block or two together."

"What did you call each other?"

"I called him Mr. Einstein. He called me Miss Hinds. Then he started calling me Violet."

She moved to Philadelphia; that's why they stopped seeing each other. When my mother died, we had a memorial service at the Episcopalian Church. Uncle Albert [Hinds] spoke there. He said: "Violet was a—I'll use the term buddy —of Albert Einstein."

Henry Pannell: I guess everybody my age in this community remembers seeing Einstein when we were kids. He'd come by—white sneakers, no socks, a loose-fitting sweater—and he'd give nickels to us kids. I remember him coming up and sitting on my grandmother's porch and chatting with her.

My grandmother, Carrie Dans Pannell, was a founding member of the Friendship Club. She lived on Jackson Street right at the edge of the Witherspoon [black] community. Like if Einstein was walking from his place on Mercer Street to the hospital which was and still is in the Witherspoon community, Jackson Street (it's now Paul Robeson Place) was the first street he would have come to, maybe a chance to rest before continuing on, maybe a chance to chat with friends, most likely both. My grandmother worked as a chef in the Nassau Club and was active in the Friendship Club, along with Emma Epps, Bertha Hill Brandon, and others.

Einstein used to talk to everybody in our community. He didn't just come and sit on my grandmother's porch, but on the Wilsons' porch and others. He'd talk with everyone.

As a boy, I always felt comfortable around him. I never really talked to Einstein, but I remember him walking in the community when I was a kid. And I was really great friends with his secretary, Helen Dukas. She was a wonderful lady.

When I worked at the Institute for Advanced Study, the print shop was in the basement, and on the left was this huge room with a safe door on it—it was sort of a vault. Miss Dukas worked in that air-conditioned, heat-controlled room. She was working on Einstein's papers—all his papers were in there. This was after he had passed away, and they knew how valuable all his papers were so they kept them down there in that safe or vault. The door was a couple of foot thick—steel—safe door. At the time, Miss Dukas's eyesight was failing and she trusted me with the combination. Nobody else knew it, but I used to open the safe for her, because she couldn't do the combination. She treated me like one human being to another human being.

McCarter Theatre is the preeminent concert hall in Princeton. It was there that Einstein saw a 1937 performance by the great diva Marian Anderson. Born into a poor South Philadelphia family, Anderson became one of the greatest voices in opera, but racism and segregation in America denied her a wide audience. When she was six, she sang in the choir at the Union Baptist Church, but she did not take her first formal music lessons until she was fifteen, after her church raised the money for her. Like many black performers of that era, she could not build a reputation at home until she had performed abroad. So in 1930, the great contralto went overseas on a fellowship and soon took Europe by storm, drawing crowds and rave reviews in cities across the continent. The famous conductor Arturo Toscanini told her, "A voice like yours is heard once in a hundred years."

Princeton Group Arts, an organization that provided African American youngsters with art instruction not available in their segregated Princeton school, sponsored Marian Anderson's performance at McCarter. The concert was the brainchild of the Group's leader, noted African American artist Rex Goreleigh.[1]

Not surprisingly, her scintillating performance drew high praise in the Princeton press ("complete artistic mastery of a magnificent voice"). However, despite the accolades Anderson received, her international fame, and an overflow audience at McCarter, the African American contralto was denied a room at Princeton's whites-only Nassau Inn. Albert Einstein promptly invited her to stay with him, Margot, and Helen Dukas (Elsa had passed away in 1936). The diva accepted Einstein's offer and their friendship continued for the rest of his life. Whenever she returned to Princeton, Marian Anderson stayed at Einstein's house on Mercer Street.[2]

> *Albert Hinds*: I remember Marian Anderson's concert at McCarter Hall. I was involved in that event—when she was refused a room at the Nassau Inn and Einstein invited her to stay at his house with him and his daughter, Margot.
>
> I was working with Rex Goreleigh and Spring Street Group Arts—Goreleigh was director. Group Arts brought lots of African American artists to Princeton. They brought Carmen McCrae and Duke Ellington. But blacks did not stay in the hotels; most went to Mrs. Hill's house and another rooming house, Mrs. Dickerson's, also on Green Street.

In 1939, the Daughters of the American Revolution barred Marian Anderson from singing at their Constitution Hall in Washington, D.C. Instead she gave a concert, arranged by Eleanor Roosevelt, at the Lincoln Memorial before more than 75,000 cheering people.

Despite her worldwide renown, it was not until January 1955 that Anderson was finally permitted to sing with New York's Metropolitan Opera. That same month, Princeton's Friendship Club brought her back for another concert at McCarter. The opera star decided to stay with her old friend Albert Einstein who had only a few months to live. When she left, she later wrote, "I knew this was really good-bye."[3]

> *Alice Satterfield*: I worked in the kitchen at the Institute for Advanced Study in the early forties for three to five years, and got to know Einstein. Shirley was in school and I would be leaving the Institute to walk home. If I didn't get the bus and Professor Einstein would be walking, we'd walk together until we got to his residence. I'd bid him good-bye and continue home.
>
> We didn't talk a lot—on a couple of occasions he held my hand without saying anything. He would just walk in a silent and wonderful way in which you knew everything would be all right.
>
> You felt good walking with him. He did not look down on people. He was inspirational.

Some of Ms. Satterfield's friends say she is pretty inspirational herself. When she was younger, she decided to sit in the whites-only section of Princeton's movie house, the Garden Theatre.

> *Alice Satterfield*: I just felt I'm gonna do it anyway. We talked about breaking the taboos of racist seating and such, so I just sat where I wanted to in the Garden Theatre. I remember it was Friday afternoon, they had programs for kids and that day it was *Rin Tin Tin*, the movie dog. Colored were supposed to sit on one side. Instead, I sat in the white section. Nothing happened to me—maybe I was lucky.
>
> I remember not being allowed into Balt Bakery. We weren't supposed to go in. Heck, there were lots of other stores on Nassau Street that were segregated, too. I didn't like that kind of thing, so I'd go into Balt anyway.
>
> The Institute for Advanced Study had its own private bus that brought the workers, black and white, back to town. I

often took that bus and sometimes Einstein was on the same bus. He was a man of great warmth. Too bad there aren't more people like him.

On those buses, you could sit where you wanted to sit. The Institute itself was filled with people from all over the world. A gentleman from India and I would sit together and ride until we got to Nassau Street. The Institute was integrated as far as foreigners and everyone; it was different from the university, which was segregated.

Still, the Institute wasn't all that integrated. There was a definite class structure—clearly they didn't want people to mingle with the help. Dr. [J. Robert] Oppenheimer would have a big party every year—I remember one professor got reprimanded for dancing with me. He said to me, "I don't understand why I cannot dance with you." I don't recall Einstein going to these parties.

Ms. Satterfield's story is not the only one about Dr. Oppenheimer and the Institute frowning on fraternization between its scientists and those in "lower-level" positions (usually nonwhites). "I can tell you a true story about Griggs [black-owned] restaurant in the late 1940s," Freeman Dyson, the world-renowned theoretical physicist at the Institute, recalled:

> I ate supper there regularly with David Bohm* who was then at Princeton University. The food was good and cheap. Ham and cabbage was my favorite dish. I received a letter from Kay Russell, who was Oppenheimer's secretary, saying it was inappropriate for a member of the Institute to be eating at Griggs. Of course, I ignored the letter and continued eating at Griggs. I never found out whether Oppenheimer instigated the letter or whether he knew about it. I

*Bohm, a brilliant physicist, was fired by Princeton president Harold Dodds in 1950 after he refused to name names before Senator Joseph McCarthy's inquisitors. Dodds's well-known anti-Semitism might well have been a factor in the ouster of Bohm, who was Jewish.

rather suspect that Kay would not have sent it without his approval.[4]

Shirley Satterfield: Einstein didn't look down on people. I would often be with my mother in the Institute kitchen. Einstein would come in at lunchtime. . . . I remember this man with all of this white hair. He was so nice. He'd take me for walks. I remember playing on the balcony and walking around the Institute with Einstein. I'd also go in his office with him. I'd say I was about six years old.

I just remember he was very kind and his hair was always, well, like he never combed his hair. He would take me by the hand, walking with a cane in his other hand, and talk to me. I can't remember anything of what he said, but I remember that accent and that he spoke really soft. He looked different because he had all that hair and wore sandals in all weather.

He used to eat raw eggs. And I used to eat all the olives—they had big jars of olives in the Institute kitchen. They also fixed a bread pudding in that kitchen that I loved—I always used to eat in the kitchen and the cafeteria. My mother worked real hard, but when I came home, I would talk to my family. All of them took care of me. We lived with my grandmother and my three uncles who were all in the war [World War II]. If my mother was out working, my aunt or grandma took care of me. Everybody knew everybody then. It was a close-knit neighborhood.

We used to call going to Nassau Street "going Uptown." It was a big thing to go Uptown, but there were certain places we couldn't go in. Like a restaurant called Lahiere's.

Albert Hinds: There was also a restaurant—Renwick's on Nassau Street—that barred black folks from eating there, even in the 1950s.

Shirley Satterfield: And the Balt bakery, like my mother said, we weren't allowed in there, either. Besides the Garden movie theater, they had the Playhouse. It was so beautiful.

The Playhouse backed onto Jackson Street where the black people lived. (Now it's Paul Robeson Place.) We used to go and sit there all Saturday, and watch newsreels and cartoons and movies. I used to wonder why we always sat in the back on the right-hand side. I thought that was just where we always sat, but actually that's the only place we *could* sit.

Einstein: Light to the Power of 2, a one-hour television show, was written, produced, and directed by Raymond Storey, Richard Mozer, and David Devine in 1997. Though they did not base their story on real people, nor had they done any research in Princeton, much of what they produced eerily echoes the truth.

Einstein meets a young African American girl named Lannie Willis, who is doing her homework on Princeton's campus. The scientist gains Lannie's trust by rebuffing two white college students who (in a scene full of racial implications) roughly tell her she has no business on the campus because it's private property. Later, in exchange for some gumdrops, Einstein agrees to help Lannie with her homework. She and Einstein become friends.

We find that Lannie is the only black kid in her class. Her white teacher is negative, condescending, and makes comments like "you belong in vocational school." She wants to send Lannie to a remedial class. In a meeting with the teacher, Lannie's mother, realizing that the remedial class is nothing but a dumping ground for black students, angrily rejects her suggestion. Lannie remains in the classroom but her confidence is shattered. Einstein encourages her—by recounting how he did not like school when he was her age and by insisting that she's just as smart as everybody else.

The movie climaxes at a science fair in which Lannie is to make a critical (in terms of her grade) presentation. Einstein makes an unannounced visit and Lannie's parents are pleasantly surprised. The world-famous scientist walks into the classroom first, and Lannie's flabbergasted, star-struck, teacher Miss Fitch immediately recognizes him. "Wow, Dr. Einstein, I'm so happy to see you. I teach the fourth grade!" "Oh, so you must be Miss

Fitch," Einstein replies. She is thrilled that he knows her name. He asks if she can take his coat. She happily does and Einstein says, "and yes you can also take," he points to Lannie's parents, "their coats as well. We're all together." Lannie, of course, makes an A+ presentation. Like the typical TV ending, things wind up happily, but the point about racism is cogently made.

"It really sounds like the way the schools were here," says Shirley Satterfield, speaking about her educational experience in Princeton. The New Jersey legislature passed a law in 1947 that made segregation in public schools illegal. "I believe Princeton was the last town to integrate educationally in New Jersey," says physicist Freeman Dyson, whose children began attending public school there in 1948. The Princeton Plan for School Reorganization also began in 1948. The Princeton Plan was quite simple. The all-white Nassau Street School accepted all children from kindergarten through fifth grade, and all children from sixth to eighth grades went to the former Witherspoon School for Colored Children, which was renamed the John Witherspoon Middle School. Howard B. Waxwood, who had been the principal at the Witherspoon School since 1936, was retained as principal of the new middle school. He was the first and only black principal of an integrated school in Princeton until he retired in 1968.

For black Princetonians, the Princeton Plan had an unexpected consequence: it brought about the demise of the remarkable educational institution, the Witherspoon School for Colored Children. Alice Satterfield and Albert Hinds are graduates of the Witherspoon School, as is Mr. Hinds's mother, class of 1894. "We were ahead of some white kids who went to Princeton High School, as far as spelling and diction," says Satterfield. "They drummed it into your head and if you didn't get it the first year, they kept you back the second year. They taught us manners. They taught us to love." Alice Satterfield's mother Annie Van Zandt taught at Witherspoon School and one of her students was Paul Robeson. "The old school building still exists," according to Hinds. "A nursing home is now there."

Other unforeseen issues arose in the African American Witherspoon Street community as a result of the Princeton Plan. That September, black parents quickly found that their children were summarily being put in remedial classes and steered into vocational programs. Insulted and outraged by this policy, they met with public school administrators and demanded that their children receive equal treatment. Though the officials technically acceded to their request, this was the beginning of a long difficult struggle for African Americans in Princeton's public school system. "The Princeton Plan did little to alter community racial attitudes. Most black students continued to lead segregated lives within school. They were met with low expectations from teachers and discouraged from participating in extracurricular activities."[5]

Shirley Satterfield and Henry Pannell were among the first black students to experience the Princeton Plan. They were both in third grade and still bear some scars. "When they integrated schools they let us down educationally," says Pannell. "In third grade we were laughing about something," recalls Satterfield. "One of the kids said, 'Look at Shirley blush' and the teacher said, 'Shirley can't blush, she's a Negro.'"

CHAPTER 5

Einstein and Robeson, I

If Einstein had moved to Princeton a quarter-century earlier, one of the youngsters he might have met while walking along Witherspoon Street was a tall, athletic, sharp-witted preteenager named Paul Robeson. It is easy to imagine them walking and talking together, much as Einstein did several decades later with eleven-year-old Harry Morton (see chapter 11). Einstein eager to learn about his new friend's world, the young Robeson proudly talking about his oldest brother, Bill, who was studying in Philadelphia to become a doctor, and his father, who escaped from slavery in North Carolina at the age of fifteen, then helped the Union Army in the Civil War, and later became a minister and moved to Princeton to work at the Witherspoon Presbyterian Church for twenty-one years. Then the white people in charge fired him.[1]

In fact, by the time Einstein arrived in Princeton in 1933, Robeson was long gone. After moving to nearby Somerville, where his father built and led the AME Zion Church, young Robeson graduated from Somerville High School and won a scholarship to Rutgers, a private, overwhelmingly white college at the time. A straight-A student for four years, Robeson was also a member of the varsity debating team, literary society, and one of only eight Rutgers students elected to Phi Beta Kappa in both his junior and senior years. He also earned an astonishing fifteen varsity athletic letters. But his most famous achievement came in football where, despite initial racist brutality against him from teammates,* "Robeson of Rutgers" became a gridiron

*"On the first day of scrimmage they set about making sure that I wouldn't get on their team. One boy slugged me in the face and smashed

legend.* The *New York Tribune*'s report on Rutgers' game against a heavily favored Naval Reserve team (boasting All-American players from a dozen top colleges) almost sounds like an allegory about Robeson's later life:

> A tall, tapering Negro in a faded crimson sweater, moleskins and a pair of maroon socks ranged hither and yon on a wind-whipped Flatbush field yesterday afternoon. He rode the wings of the frigid breezes: a grim, silent, and compelling figure. Whether it was Charlie Barrett of old Cornell and All-American fame or Gerriah or Gardner who tried to hurl himself through the [line], he was met and stopped by this blaze of red and black.
>
> The Negro was Paul Robeson of Rutgers College, and he is a minister's son. He is also nineteen years of age and weighs two hundred pounds. . . . It was Robeson, a veritable Othello of battle, who led the dashing little Rutgers eleven to a 14–0 victory over the widely heralded Newport Naval Reserves.[2]

Ignoring his father's advice that he become a minister, Robeson earned a law degree from Columbia, helping to pay the tuition by playing professional football between semesters. (He had also been accepted by Harvard Law School, but chose Columbia because he wanted to live in Harlem where the Harlem

---

my nose. That's been a trouble to me as a singer every day since. And then when I was down, flat on my back, another boy got me with his knee, just came over and fell on me. He managed to dislocate my right shoulder." See "Robeson Remembers—An Interview with the Star of Othello," by Robert Van Gelder, *New York Times*, January 16, 1944.

*Although Robeson was named an All-American, his name was later deleted from the official college football record book, which for years listed only ten names on the 1918 All-American team (and the 1917 nationally picked team). He was denied his place in the College Football Hall of Fame until 1995. Robeson was also omitted from Rutgers University's 1954 list of its top sixty-five athletes.

Renaissance* was in its early stages.) Racism encountered in his first law office quickly helped Robeson decide to change careers, and he went on to become a world-famous singer and actor (and, later, a political activist). By the mid-1930s, he had won critical acclaim for performances throughout the United States and Europe, including his stage portrayals of Eugene O'Neill's *The Emperor Jones* and, in London, Shakespeare's *Othello*.

Robeson's attitude toward Princeton as he traveled the world was similar, in many respects, to the way the young Einstein had viewed his boyhood home. Einstein had escaped from German militarism and anti-Semitism, while Robeson described the Princeton he left behind as "a Georgia plantation town," and "spiritually located in Dixie."[3]

But for Robeson there were two Princetons. The African Americans in the Witherspoon community where he was raised were "hard-working people, and poor, most of them, in worldly goods—but how rich in compassion!" They lived, he said, "a much more communal life than the white people around them," and he would not forget his roots: "I had the closest of ties with [Princeton's black] workers since many of my father's relatives—Uncle Ben, Uncle John, Cousin Carraway, Cousin Chance and others—had come to this town and found employment at such jobs—domestics in the homes of the wealthy, cooks, waiters and caretakers at the university, coachmen for the town and laborers at the nearby farms and brickyards."[4] When Robeson was at Rutgers and Columbia, "he used to come back and sing for us at church concerts," according to one member of Princeton's Witherspoon Church congregation.[5]

It is not surprising, then, that in October 1935, having just returned to America from another successful stint on the European stage and on his way to Hollywood for a highly publicized and lucrative filming of *The Emperor Jones*—with concert stops

*The years from 1917 to 1935 witnessed a flowering of works by African American poets, playwrights, artists, and writers known as the Harlem Renaissance. Harlem at that time was the cultural capital of black America.

1. Lillie Trotman. *Photograph by Mary Trotman.*
2. From left to right, Shirley Satterfield, Penney Edwards-Carter, and Henry Pannell. *Photograph courtesy of Henry Pannell.*
3. Mercedes Woods (seated). *Photograph by Henry Pannell.*
4. Wallace Holland. *Photograph by Henry Pannell.*

5

6

7

8

9

10

11

12

13

5. Harriet Calloway. *Photograph by Henry Pannell.*
6. Consuela Campbell. *Photograph by Henry Pannell.*
7. Alice Satterfield. *Photograph courtesy of Shirley Satterfield.*
8. Lloyd Banks. *Photograph by Henry Pannell.*
9. Morris Boyd. *Photograph by Henry Pannell.*
10. Mary Trotman. *Photograph courtesy of Mary Trotman.*
11. Violet Hinds Jones (seated) with her daughter Shirl Gadson (left), her granddaughter and great granddaughter. *Photograph courtesy of Shirl Gadson.*
12. Alice Satterfield, second from right, standing, with her colleagues outside Fuld Hall at the Institute for Advanced Study in 1948. *Photograph courtesy of Shirley Satterfield.*
13. Timmy Hinds. *Photograph by Henry Pannell.*

14. Joi and Harry Morton. *Photograph courtesy of Joi Morton.*
15. Rod Pannell. *Photograph by Henry Pannell.*
16. Fanny and Jim Floyd. *Photograph courtesy of Fanny Floyd.*
17. Albert Hinds at age 102. *Photograph by Fred Jerome.*

18

19

18. Witherspoon Street Presbyterian Church.
Photograph by Claude Satterfield, courtesy of Shirley Satterfield.
19. Students (including Paul Robeson) at the Witherspoon School, circa 1903. *Photograph courtesy of Shirley Satterfield.*

20

21

22

20. Witherspoon School. *Photograph courtesy of Shirley Satterfield.*
21. Street sign at the intersection of Witherspoon Street and Paul Robeson Place in Princeton, 2004. *Photograph by Henry Pannell, with thanks to Wayne Carr, Bill Euricun, and Jason Morgan.*
22. Robeson as Othello in a production at Princeton's McCarter Theatre in August 1942. *McCarter Theatre Records, Box 2, Princeton University Archive. Photograph courtesy of the Department of Rare Books and Special Collections, Princeton University Library.*

23

24

23. Einstein lecturing on relativity to Lincoln University students, May 3, 1946. *Photograph courtesy Lincoln University of Pennsylvania Archives.*
24. Einstein receiving an honorary degree (D.Sc.) from Dr. Horace Mann Bond, president of Lincoln University, at the Conference on Objectives, May 3, 1946. *Photograph courtesy of Lincoln University of Pennsylvania Archives.*

25

26

25. Einstein with children of Lincoln University faculty members. From left to right, Larry Foster, Stephanie Reynolds, and Yvonne Foster. *Photograph courtesy of Yvonne Foster Southerland.*

26. From left to right, Henry Wallace, Albert Einstein, radio commentator Frank Kingdon, and Paul Robeson at Einstein's Princeton home in October 1947. *Photograph © Bettmann/CORBIS.*

en route in New York, Chicago, Milwaukee, Portland, and Seattle—the now-renowned Robeson found time or made time to travel to Princeton for a concert at the McCarter Theatre to benefit the black YMCA. Located in a small, wood-frame building on Witherspoon Street, the Y was forever desperate to fund programs in the arts, sports, and education for the children in Robeson's old neighborhood.

Einstein had most likely read about Robeson before they eventually met in Princeton. Their paths first crossed in Berlin in 1930, where Robeson gave two performances of *The Emperor Jones.* The Nazis had not yet taken power, although armed fascist gangs were increasingly violent in the streets of Germany and across Europe. Robeson had recently concluded a successful concert tour of European cities where, according to one report, his "tumultuous, unprecedented receptions became anti-fascist demonstrations."[6] In Berlin, the German media, especially the Social Democratic press, which Einstein read regularly, hailed Robeson in *The Emperor Jones* (even as they disparaged O'Neill's play).[7] While he missed seeing the play, Einstein could hardly avoid the rave reviews and Robeson's photos. The most obvious explanation for Einstein's passing up *The Emperor Jones* in 1930 is simply that his English was not good enough. Ten years later he would see the play in Princeton. By then he would know English better—and he would know Robeson.

Einstein and Robeson may not have met each other face-to-face in Berlin, but they both met the face of "New Germany." Einstein had frequently endured anti-Semitic attacks in the media, crank letters, and death threats. But more terrifying was the scene of May 10, 1933, in the Unter den Linden square facing Berlin's opera house. Though he no longer lived in Germany, Einstein heard enough vivid accounts of the event to feel the chill: some five thousand frenzied, swastika-wearing Nazi youth tossed thousands of books into bonfires, shouting "*Brenne* [burn] Karl Marx! *Brenne* Freud!" They also burned the works of

Thomas Mann, Erich Maria Remarque, André Gide, Helen Keller, Upton Sinclair, and other Nazi-designated "Communist literature," including books on the "Jewish theory" of relativity by Albert Einstein. A crowd of some forty thousand Berliners stood and watched—and cheered. William Shirer later called it "a scene that has not been witnessed in the Western world since the late Middle Ages."[8]

For Robeson, the "New Germany" appeared on a platform at Berlin's central railroad station a year and a half later, on December 21, 1934, as he stood with Marie Seton, biographer of the Soviet filmmaker Sergei Eisenstein, while his wife, Eslanda, went to check on the luggage. The three, on their way to Moscow where Robeson planned to make a film, had stopped in Berlin to change trains. When an elderly German woman saw the black man and white woman together, she stopped staring just long enough to report her sighting to three men in Nazi uniforms who turned to look in Robeson's direction. "I could read the hatred in their eyes," he told a German newspaper many years later. It reminded him of a lynch mob, and he remembered his older brother Reeve telling him, "If you have to go, take one with you." "I took a step forward, and they could read something in my eyes." The uniformed men walked away, and, when Essie returned, the three travelers boarded their train to Moscow. "For a long time after the train moved out of Berlin," Seton later wrote, "Paul sat hunched in the corner of the compartment, staring out into the darkness."[9]

Their first certain meeting came on October 31, 1935, in Princeton, when Einstein went backstage after Robeson's concert at the McCarter Theatre. It is easy to envision Helen Dukas, Einstein's longtime assistant, returning from shopping with the *Daily Princetonian* and showing Einstein, Elsa, and Margot the front-page story about the performer they had missed seeing in Berlin, who would be singing that evening in Princeton. The article, surrounding a handsome, smiling photo, ran under the headline:

ROBESON, BRILLIANT BARITONE AND ACTOR, WILL PRESENT SPIRITUALS IN HOME-COMING

One can imagine, too, the household gathered in the living room as Helen reads aloud to the others, translating some words into German:

> Baritone, actor, scholar, athlete and lawyer, Paul Robeson of Princeton returns to his birthplace tonight, when he will sing spirituals on the McCarter Theatre stage at 8:15 in his only New Jersey appearance of the year.
>
> The proceeds of the concert are for the benefit of the Witherspoon YMCA. Tickets are priced from $2.20 to $1.10. . . .
>
> Winning at Rutgers letters in football, baseball, basketball and track, a Phi Beta Kappa key, a 90 per cent scholastic average and the reputations of being "the greatest defensive end that ever trod the gridiron" and "the perfect type of college man," Robeson continued his career with a law course at Columbia.
>
> . . . From the days of "Emperor Jones" to the present time, Paul Robeson has performed before packed houses in all the large cities of this country and in the capitals of Europe. . . . He created a sensation as the Moor in Shakespeare's "Othello" in a London production in the spring of 1930, when English and New York critics filled their papers with columns of praise. Robeson was the first Negro to act in the role since the days of Ira Aldridge 49 years ago. . . .
>
> It is in the rhythm, exaltation, emotion, simplicity and beauty of his great voice that Robeson attains the high degree of perfection which is felt when he sings such spirituals and Negro songs as "Old Man River," "Deep River," "Water Boy" and "Were You There?"

Later, backstage after the concert, Einstein, still energetic at fifty-six, reached out to shake Robeson's hand. The singer, taller, broader, nineteen years younger, and celebrated around the world, was awed. "I am honored," he began in his deep, gentle voice, but Einstein interrupted: "No, it is I who feel honored." According to his friend and colleague Lloyd Brown, years later, when Robeson described the meeting, the awe was still in his voice. Besides the evening's concert, the two men discussed their mutual alarm at recent world events. Within the prior six weeks, the Nazi regime in Germany had issued the Nuremberg Laws, canceling citizenship for Jews and outlawing marriage or sexual relations between Jews and non-Jews, and Mussolini's troops had brought Italian fascism to Africa by invading Ethiopia. In their brief, first meeting, Einstein and Robeson discovered they shared not only a passion for music but a hatred of fascism.[10] Within a year, both men joined the international effort to defend the democratically elected government of Spain from Francisco Franco's fascist armies.

The *Princetonian*'s laudatory front-page prewrite for Robeson's concert indicated his phenomenal popularity in 1935, but the reasons for that popularity were made clearer in the next morning's review:

> PAUL ROBESON THRILLS MCCARTER AUDIENCE WITH RICH TONES AND WARM PERSONALITY
>
> A rendering of Negro songs, both spiritual and secular, better than any of those with which Paul Robeson thrilled his audience in last night's magnificent concert in McCarter would have been hard to find. Though the effects of a cold were plainly audible in his voice they were quickly forgotten when the deep richness of his tones and his irresistible personality took hold of the listeners. . . .
>
> What gives to Robeson's singing its peculiar potency is the amazing extent to which he is able to live what he sings, to put himself into his songs. In the best numbers, he seems actually to become the song, his whole body reacts to it, he walks about the stage, taps his foot in time to the rhythm and in still other ways expresses the composition. Yet he never comes near being sentimental and brings a true greatness of interpretation to his more serious numbers. The spell which he cast over the audience in "Nobody

> Knows the Trouble I've Seen," "Go Down, Moses," and to some extent in "Water Boy" was indeed a profound one. The first-mentioned in particular made a most moving impression upon this reviewer.[11]

If he had seen the Einstein-Robeson meeting after the concert, the *Princetonian* reviewer might have added a footnote to his story. But to avoid curiosity seekers, Princeton's most recognizable resident probably waited until almost everyone else had gone before making his way backstage. In fact, whenever Einstein attended McCarter, the theater management arranged "to sneak him into the audience fifteen minutes early and keep the lights dimmed" so he wouldn't be "battered by admirers and newsmen."[12]

Einstein and Robeson met for a second time during the week of August 12–17, 1940, when Robeson played *The Emperor Jones* at McCarter, and again at McCarter in August 1942, when Robeson did *Othello* with Uta Hagen and Jose Ferrer.[13]

For Princeton, the McCarter Theatre was one of those strange contradictions that appear occasionally in America's racial attitudes. It was similar to the paradox of millions of white people hailing Jackie Robinson's entry into white baseball in 1946,[14] while, during the same year, they shrugged as scores of black Americans were lynched. Here was a theater in a town modeled on Mississippi's values, a town that didn't enter the twentieth century until it was half over, dominated by a university whose decision-makers kept their white hands clutching the old supremacy myth, and kept their schools, stores, and movie theaters segregated well into the 1940s. Yet because McCarter was privately owned, it was not bound by the biases of Princeton—town or gown—and, seeking financially successful events, could host Robeson and Marian Anderson in sold-out, critically acclaimed performances before standing-room-only—racially integrated—audiences.[15]

Einstein could not have known it when he visited backstage,

but for Robeson there was a special irony in performing *Othello* in Princeton—after Broadway theaters had already rejected it as too racially risky. Margaret Webster, the British producer, reported that every New York theater manager she approached was afraid of a production in which a black man made love to a white woman: "Mostly they were just plain scared of the issues which the production would raise."[16] When Webster and Robeson decided they'd try to open the play in a theater outside New York, they found only two who agreed to take it: the Brattle Theatre in Cambridge, Massachusetts, and the McCarter. So the kiss that couldn't play on Broadway succeeded in the same town whose high school and university had barred Robeson and his brothers, the same town where Reverend Robeson's militancy had cost him his church. A year after its out-of-town opening won rave reviews, *Othello* opened in New York at the Shubert Theatre for a run of 296 performances, a record for a Shakespeare production on Broadway.

For Robeson, the taste of irony may have been bitter-sweet, given the Jim Crow conditions his cousins, friends, and others still faced in Princeton's black community. "It means so little when a man like me wins some success," he told a reporter from Wisconsin. "Where is the benefit when a small class of Negroes makes money and can live well? It may all be encouraging, but it has no deeper significance. I feel this way because I have cousins who can neither read nor write. I have had a chance. They have not. That is the only difference."[17]

And contradiction or not, Princeton was still Princeton: a black man whose mother "worked in several of the homes in the university" reported that when *Othello* came to town, his mother overheard at parties "much discussion about the fact that in the McCarter Theatre Paul Robeson kissed a white woman. . . . It was a play but they could not accept that, and they showed their southern upbringing and their southern attitudes."[18]

To Alice Satterfield and her daughter Shirley, Paul Robeson was almost part of the family: "My grandparents Wayman and

Martha Van Zandt lived on Edgehill Street," Alice says. "My granddaddy had his own house. My mother Annie was a schoolteacher. She taught at Witherspoon School for Colored Children and one of her students was Paul Robeson. Paul's father was minister of our church for twenty-one years. That was a time of deep Jim Crow-ism and his father was always fighting for the rights of Negroes."

Shirley recalls that after Paul Robeson left Princeton,

> he used to come back to see my grandma Annie, and when he came, he used to sit me on his lap and tell me stories. I remember his deep voice. He used to come back and have concerts at McCarter Theatre. He was a loved man, but I think he became bitter because Princeton was a Jim Crow town. He loved the people in the community, he just questioned how people were treated. He left town because of the way he was treated.

When Robeson played *Othello* at the McCarter Theatre, Shirley recalls,

> He was wonderful with that big booming voice. Paul would come to Princeton because his relatives lived on Green Street and he would make the rounds. He would always be walking, he'd stop at our house, he would go on down to Birch Avenue. As kids, we'd heard about him, and we would run behind him to the corner—we had all heard about him, we knew he was someone great. Then when he played *Othello* we went to see it and we were just thrilled.

There must have a sense of reunion in the air when Einstein came backstage after Robeson's *Othello* performance in mid-August 1942—Robeson anticipating the arrival of the great professor, now his "old friend" and regular backstage visitor in Princeton, and Einstein eagerly looking forward to another chat with the warm and brilliant star who shared so much of his worldview. Perhaps they embraced when they met. This reunion, however, had a somber tone. The antifascist struggle, so central in each of their lives, had failed to stop Franco in Spain, and now there was a world war.

Less than a year after their first backstage meeting in October 1935, when Einstein and Robeson had shared their concern about the rising fascist tide in Europe, Francisco Franco's troops, with planes and tanks provided by Hitler and Mussolini, attacked the popular-front government of Spain, which included socialists and communists. For the next three years, support for the Spanish Republic became a central cause for both Robeson and Einstein, a cause that brought them into conflict with the Roosevelt administration's foreign policy of neutrality. Washington's neutrality barred sending arms or troops to either side but permitted Texaco to sell oil to Germany and Italy—oil that fueled the planes dropping bombs on hundreds of Spanish towns.

One of those bombing raids became a symbol for a new kind of war, in which men miles high in the sky drop death and terror on civilians they cannot—and do not want to—see.

April 26, 1937, was market day in the small Basque town of Guernica, where some ten thousand peasants and others from the surrounding region crowded into and around the central plaza. In mid-afternoon, a squadron of Heinkel 51 German airplanes suddenly appeared overhead and, at 4:40 p.m., they began bombing the town. The bombing continued for three hours, "slaughtering women carrying their babies, shop-owners protecting their goods, and peasants running for cover in the fields.* When the dust had settled, more than sixteen hundred people were dead and eight hundred others wounded." The Germans would later claim it was a military strike aimed at the

*Picasso's *Guernica*, created in 1937, is not only the best-known, but perhaps the world's most chilling antiwar or antifascist work of art. After hanging in New York's Museum of Modern Art for decades, it was returned to Spain in 1981 after the death of Franco, in accordance with Picasso's wishes. It now is exhibited in the Reina Sofia National Museum in Madrid.

town's single munitions plant, but, "in fact, just about the only parts of Guernica not touched by destruction were the munitions factory and the bridge."[19]

Some 35,000 volunteers from fifty-two countries traveled to Spain to help resist the fascist assault. The international volunteers included some 2,800 American men in the Abraham Lincoln Brigade:*

> The Lincolns came from all walks of life, all regions of the country, and included seamen, students, the unemployed, miners, fur workers, lumberjacks, teachers, salesmen, athletes, dancers, and artists. They established the first racially integrated military unit in US history and were the first to be led by a black commander . . . Oliver Law, a 33-year-old army veteran from Chicago.[20]

Under the banner of neutrality, the Roosevelt administration had banned travel to Spain, but the Lincolns managed to get there anyway, most crossing the Pyrenees from France on foot with the help of French antifascists. Nearly 750 Lincolns were killed in battle, with hundreds more wounded.

Few twentieth-century conflicts polarized right and left as sharply and publicly as the Spanish Civil War. Conservative groups consistently supported Franco, even after World War II. In 1945, for example, Mississippi's congressman John Rankin attacked Einstein for endorsing the American Committee for Spanish Freedom (RANKIN WANTS FBI TO CURB EINSTEIN, declared a *Detroit News* headline). The House Un-American Activities Committee (HUAC)—before, during, and after the Second World War—condemned Einstein as "an endorser of the North American Committee to Aid Spanish Democracy [and] a sponsor of [its] Medical Bureau," which HUAC and Hoover called

*Some eighty-five American women volunteered in the Loyalist cause. The story is told in Julia Newman's award-winning documentary film, *Into the Fire* (Exemplary Films, 2002).

"Communist fronts."[21] Others who continued for years to hail Franco and red-bait his opponents included the officials of the Roman Catholic Church and the American Legion, which gave its 1951 Award of Merit to Generalissimo Francisco Franco.

Volunteer armies rarely, if ever, organize themselves, and in the case of Spain, it was no secret that the International Brigades were organized by Communist parties around the world.[22] But support for the beleaguered Spanish Republic extended far beyond the left, involving liberals, democratic socialists, even anti-communists (in England, Clement Attlee, head of the British Labor Party and future prime minister, endorsed the anti-Franco campaign), and an ocean of ordinary people who never wore a political label. Defending Spanish democracy became an international rallying cry in dozens of languages and the theme of books, poems, songs, and even Hollywood movies. A broad array of artists, scientists,[23] and activists joined the cause, including Americans such as Ernest Hemingway, Langston Hughes, Helen Keller, Gene Kelly, Gypsy Rose Lee, Dorothy Parker, and A. Philip Randolph—as well as Einstein and Robeson.

A number of African American intellectuals and artists adopted the Spanish cause as their own. Black newspapers, including the *Pittsburgh Courier*, the *Baltimore Afro-American*, the *Atlanta Daily World*, and the *Chicago Defender*, unequivocally sided with the Spanish Republic and occasionally carried feature articles about black participants in the Lincoln Brigade. Several black medical personnel from the United Aid for Ethiopia (UAE) offered medical supplies and raised money in the community; Harlem churches and professional organizations sponsored rallies in behalf of the Spanish Republic; black relief workers and doctors raised enough money to purchase a fully equipped ambulance for use in Spain; and some of Harlem's greatest musicians, including Cab Calloway, Fats Waller, Count Basie, W. C. Handy, Jimmy Lunceford, Noble Sissle, and Eubie Blake, gave benefit concerts sponsored by the Harlem Musicians' Committee for Spanish Democracy and the Spanish Children's Milk Fund.[24]

Robeson and his wife, Eslanda, went to Spain during the last week of January 1938. For months, Robeson had been singing and speaking at concerts and mass meetings in London (where the family was living at the time) to rally support and raise money for the Spanish Republic. At one meeting in the Albert Hall in December, Robeson told the overflow crowd, "The artist must take sides. We must elect to fight for freedom or slavery. I have made my choice." In words that Einstein might have echoed, Robeson said that scientists, too, had to choose: "Every artist, every scientist, must decide now where he stands. . . . Through the destruction—in certain countries—of the greatest of man's literary heritages, through the propagation of false ideas of racial and national superiority, the artist, the scientist, the writer is challenged. The battlefront is everywhere."[25] A month later, he and Essie were in Barcelona.

Robeson's first interview in Spain was with the Afro-Cuban poet Nicolas Guillén who asked why he had decided to come to the war-torn country: "It is not only as an artist that I love the cause of democracy in Spain," he told Guillén, "but also as a black man. I belong to an oppressed race, discriminated against, one that could not live if fascism triumphed in the world. My father was a slave, and I do not want my children to become slaves."[26]

For a week, the Robesons traveled across what was left of Republican Spain. They witnessed the devastation from fascist bombings of civilian neighborhoods, schools, and hospitals, and Robeson chatted with Spanish Loyalist soldiers and antifascist volunteers from around the world—including a group of black Lincoln Brigade members from the United States.

But by far the high point of the week was Robeson's concert for the troops. He sang, in English and Spanish, new, anti-Franco songs such as "Los Quatros Generales" ("The Four Insurgent Generals") as well as old favorites like "Old Man River" with new words Robeson had recently added. Instead of the original "tired of livin' but scared of dyin'," he sang to a deluge of cheers: "I must keep fighting until I'm dying." Describing Robeson's impact on the troops, British volunteers later recalled that at first most of the men didn't believe the rumor that he was really in Spain: "You don't get people like that every day of the week running into a war to see how things are going." When they met him, when they heard him sing, "the whole place lit up . . . it was just like a magnet drawing you . . . as if somebody was reaching out to grasp you and draw you in." They felt they had been with "a friend of lifelong standing."[27]

Robeson called the trip to Spain "a major turning point" in his life. "I have never met such courage in a people," he later told a reporter. At the same time, by 1938, the outlook for the Republic was grim, and if Robeson didn't feel the impending defeat, what he had seen was not encouraging. He attacked the failure of the United States, England, and France to support the Republic. In his notebook, he wrote: "We are certainly not doing anywhere near enough. We don't feel deeply enough."[28]

Einstein expressed his anger at America's neutrality toward the fascist attack on Spain early in February 1937, declaring he felt "ashamed that the Democratic nations had failed to support the Loyalist Government of Spain." In a message to "a prominent Spanish personality whose identity was not revealed," according to a *New York Times* report, Einstein said, "I . . . assure you how intimately united I feel with the Loyal forces and with their heroic struggle in this great crisis of your country." Perhaps because it criticized a U.S. government policy, it was a statement that Hoover's FBI was fond of repeating in Einstein's dossier, as evidence of the scientist's "subversive" character.[29]

Two months later, Einstein sent a message to a New York mass meeting (also addressed by the writer Thomas Mann), explaining that poor health kept him from attending in person:

> I view vigorous action to save freedom in Spain as the inescapable duty of all true democrats. Such a duty would also exist even if the Spanish Government and the Spanish people had not given such admirable proof of their courage and heroism. The loss of political freedom in Spain would seriously endanger political freedom in France, the birthplace of human rights. May you succeed in rousing the public.[30]

Einstein also supported the Canadians who fought against Franco, sponsoring the Friends of the Mackenzie-Papineau Battalion Rehabilitation Fund, which gave assistance to Canadians wounded in Spain.

In a last-ditch effort to stop the fascist takeover, Einstein and a group of Princeton University professors issued a futile appeal to Roosevelt in the spring of 1938 to lift the U.S. embargo on sending weapons to the Spanish Republican government. In March 1939, Madrid fell to the fascists.

Nearly twenty years later, the great Spanish cellist Pablo Casals said he was "perpetually grateful" to Einstein "for his protest against the injustice to which my homeland was sacrificed." In a tribute after the scientist's death, Casals described Einstein as "a pillar of the human conscience in a time when so many civilized values seemed to be tottering."[31]

CHAPTER 6

"Wall of Fame"

By August 1942, when Einstein and Robeson met after *Othello,* events had fulfilled the prediction that Spain would be a prelude to a wider war. Their conversation that evening may have touched on Shakespeare, but it's hard to believe they didn't talk at some length about the war news. It was gloomy. Einstein's warning five years earlier that the fall of Spain "would seriously endanger political freedom in France" proved to be just the beginning. The Nazis had taken not just Paris but most of Europe, and the Wehrmacht blitzkrieg was sweeping across the Soviet Union, seemingly unstoppable. Recent reports had described the beginning of a possibly important battle for the Russian industrial city of Stalingrad, but the Nazis were predicting victory within a few weeks. Both Einstein and Robeson felt that stopping Hitler's war machine meant backing the Russians' front-line resistance and strengthening the alliance between the United States and the Soviet Union. For Einstein, that alliance was also the first step to world government—the only way to permanent peace. For Robeson, who had visited Russia, Soviet society and its planned, socialist economy seemed free of racism and represented the future. For both men, supporting the war effort had become the primary political goal.[1]

"The Good War," as Studs Terkel would later call it, was the last time—perhaps the only time—so many Americans united behind their government. For a sense of how popular the Soviet-American anti-Nazi alliance was, consider that a few weeks before seeing Einstein at McCarter Theatre, Robeson had taken part in a rally at New York's Madison Square Garden, urging America to send war relief supplies to the Russians. Other speakers at the rally included New York's mayor Fiorello La Guardia, American Federation of Labor president William Green, Einstein's friend Rabbi Stephen Wise, opera singer Jan Peerce, pianist Arthur Rubenstein and Supreme Court justice Stanley Reed.[2]

But J. Edgar Hoover and others on the far right were not impressed. Well before their 1942 *Othello* meeting, Einstein and Robeson shared another distinction, although they probably did not know it: they had become targets of Hoover and the FBI. While the war had made Washington and Moscow allies against the Axis (fascist) powers, Hoover and his congressional cohorts on HUAC seemed at least as focused on beating the Russians as the Nazis. Many of the isolationists who had opposed entering the war before the attack on Pearl Harbor in December 1941 still felt America was fighting the wrong war with the wrong ally—the Soviet Union.*

Before the war, Hoover had maintained a number of friendly ties with Hitler's police officials. Among other examples, he sent Hitler's Gestapo chief Heinrich Himmler a personal invitation to attend the 1937 World Police Conference in Montreal. The following year, he welcomed one of Himmler's top aides to the United States, and we now know that after the war he embraced "former" Nazis into his Red-hunting FBI apparatus. Until Pearl Harbor, the FBI chief also held secret talks with congressional isolationists whose campaigns received covert contributions from the German government and who did their utmost to keep the United States out of the antifascist war.[3] But from 1941 to 1945, while billboards and posters across the country urged "All Out for the War Effort" and pictured determined-looking Americans, sleeves rolled up, giving blood, joining the army, and working in hospitals or defense plants, Hoover was more discreet

*The Communist Party of the USA also opposed U.S. entry into the war and maintained a tacit alliance with the isolationists until Nazi Germany invaded the Soviet Union on June 22, 1941. At that point the CP reversed itself and threw all its energies, and those of its members, into "winning the war against fascism."

about his politics. Behind the scenes, the FBI and HUAC conducted low-profile mini witch hunts. They investigated, spied upon, and held hearings on people and groups—especially left-wing union activists, teachers, and writers—they believed to be exerting communist influence in America.[4]

During his forty-six-year reign as head of the FBI, Hoover was often described as all-powerful. Yet, for all of Hoover's power, he could not have carried out his policies without the tacit and often explicit approval of higher officials. As one leading Hoover biographer points out, Hoover's authority depended on "powerful national leaders [who] shared the FBI director's obsessive anti-Communism and yet sought to mask their own complicity and indifference to the law."[5]

FBI ears were apparently not present backstage at the McCarter when Einstein and Robeson met, but Hoover and his Bureau, as well as HUAC, had already targeted both men.* The FBI had begun intercepting Robeson's calls and letters—it would begin similar Einstein-bugging within a few years. But Hoover's attack on Einstein went beyond tapping phones and opening mail. In 1940, in an action that remained secret for nearly half a century, the FBI fed the U.S. Army a series of anti-Einstein memos ("unlikely that a man of his background . . . could . . . become a loyal American citizen"), and—based largely on those memos—Army Intelligence denied Einstein security clearance, barring him from work on the Manhattan Project that built the atomic bomb.[6]

*In 1941, HUAC's research director J. B. Matthews charged that Robeson "has made his choice for communism." Seven years later, Matthews, as "an expert witness" before a Washington State investigating committee, alleged that Einstein might be revealing nuclear secrets to the Russians. In 1952, the same Matthews compiled lists of "incriminating affiliations" of the then anticommunist NAACP and became what Schrecker describes as "a fixture on the segregation circuit," testifying before investigating committees in Arkansas, Mississippi and Florida about "Communist inroads into the civil rights movement"; see Duberman, *Paul Robeson*, 239; FBI's Einstein file, section 5, 885–890, and section 8, 15–16; and Schrecker, *Many Are the Crimes*, 393.

Despite America's wartime anti-Nazi alliance with the Soviets, Hoover's G-men prepared an extensive list of those they considered Red or pink. As *Othello* finally opened on Broadway in the fall of 1943, the FBI chief placed Robeson's name on his "Det-Com" roster—those to be put into concentration camps in the event of war with the Soviets.[7]

Ironically, during the war years, even as Hoover added both men to his lengthening list of enemies, Einstein and Robeson supported Washington's policies more vigorously than at any other time in their lives.

In 1939, after Hitler's annexation of Austria and seizure of Czechoslovakia, with a Nazi world conquest on the way to becoming reality, Einstein put aside his past differences with Roosevelt: the house was burning and the American president seemed to have the only fire truck in town. Einstein was on vacation at Peconic, Long Island, on July 15, 1939, when Leo Szilard and fellow Hungarian-Jewish refugee physicist Eugene Wigner drove out from New York City to tell him the Nazis had begun work on an atomic bomb. After discussing steps they might take, Einstein agreed to send his now-famous letter to Roosevelt (he and Szilard actually wrote it together but decided that only Einstein should sign it), urging the president to prepare for a U.S. atom-bomb program.

Eight months earlier, Hitler's SS had unleashed a ruthless pogrom against Germany's remaining Jews, assaulting, raping and killing them—nearly one hundred murdered in one night—and arresting some thirty thousand for shipment to concentration camps. The Nazis smashed so many Jewish-owned stores and synagogues that the streets were covered with broken glass. One news report described *Kristallnacht* ("night of broken glass") as "a wave of destruction, looting and incendiarism unparalleled in Germany since the Thirty Years' War . . . police confined themselves to regulating traffic and making wholesale arrests of Jews 'for their own protection.'"[8]

Building an atomic bomb was not Einstein's favorite project, but he agreed with Szilard's argument that if Hitler got the bomb first, he would hold the world hostage. Szilard later wrote that Einstein "was very quick to see the implications and perfectly willing to assume responsibility for sounding the alarm."[9]

"My dear Professor," the President wrote back to Einstein, "I want to thank you for your recent letter. . . . I found the data of such import that I have convened a board consisting of the head of the Bureau of Standards and a chosen representative of the Army and Navy to thoroughly investigate the possibilities of your suggestion regarding the element of uranium."[10] Large government projects are never so simple, and it would be two more years before the Manhattan Project was actually launched. But Einstein had decided to support Roosevelt—America seemed to be the only country with the technology to beat Hitler to the atomic punch.

Einstein became an official American on October 1, 1940. With his stepdaughter, Margot, and secretary, Helen Dukas, he took the oath of citizenship in the federal courthouse at Trenton. As the three raised their hands for the swearing-in, an alert Associated Press photographer raised his camera, despite a no-photos rule in the courtroom. Newspapers around the world printed the resulting picture—worth many a thousand words in public relations for the U.S. government.

As a prelude to becoming a citizen, Einstein had provided more "good press" for his new country. In a radio interview on the NBC network series, "I Am an American," sponsored by the U.S. Immigration and Naturalization Service, the citizen-to-be said, "In America, the development of the individual and his creative powers is possible, and that, to me, is the most valuable asset in life." And in words he quite possibly questioned a decade later during the McCarthy period, Einstein added: "In some countries, men have neither political rights nor the opportunity for free intellectual development. But for most Americans, such a situation would be intolerable." It was his warmest praise yet for the United States.

Yet even at its warmest, Einstein's support for government policies was not unconditional. He concluded his 1940 pre-citizenship radio interview by declaring that "it is all the more important . . . to see to it that these liberties are preserved and protected."[11] It was a rhetorical technique he used frequently in the next fifteen years: praising America's venerated ideals—freedom of speech and thought, equal opportunity, and equality before the law—but warning that these principles were in danger or at least needed vigilant guarding. In a little-known speech at the New York's World's Fair, shortly after he became a citizen, Einstein took a similar approach to race.

"The World of Tomorrow" was the theme of the World's Fair that opened in the spring of 1939 in New York City's Flushing Meadows (before there was a Shea Stadium) with exhibits from 58 nations, 33 U.S. states, and 1,300 private companies. Some 2,800 people a day visited the General Motors "Futurama" exhibit and were treated to a vision of "life in 1960," where technology was so efficient that everyone took two-month vacations, lived in collapsible houses, and drove (GM) cars cleanly fueled by something called "liquid air." On the way out, everyone was given a small GM pin that read: "I have seen the future." If it seemed like a great escape, perhaps that was what made it so popular to hundreds of thousands of Depression-weary fair-goers.

Yet even escape has its limits. A visitor to the fair during the first week in September remembers seeing television sets, electric washers and dryers and a refrigerator that made its own ice—"and there was a light inside of it!" But later, as she walked into the Polish pavilion, all the lights suddenly went out. At first she thought it was an electrical failure. "And then we heard over a loudspeaker that Germany had just invaded Poland, and they were closing down the pavilion. The war had started."[12]

In the fair's second year, a new exhibit opened as part of the American Common (the U.S. exhibition area). "The Wall of Fame" was singularly lacking in technological gizmos and futuristic electric miracles. It consisted simply of twenty-one panels inscribed with more than six hundred names of "immigrants, Negroes and American Indians," divided into categories of expertise, such as Government, Music, Arts, and Science, and chosen from more than six thousand names submitted to the selection committees in the various professions. The music section listed Paul Robeson, and the science section included Albert Einstein. Above the panels, in large, capitalized words, appeared the following explanation: "Inscribed here are the names of American Citizens of Foreign Birth, American Indians* and Negroes who have made Notable Contributions to our Living, Ever-Growing Democracy Devoted to Peace and Freedom."[13]

The exhibit was boldly designed to contrast with the rest of the fair, emphasizing that America's strength came not just from its technology but from its melting-pot tradition and that, in the words of the then-popular "Ballad for Americans,"† the nation is "as strong as the people who made it . . . the et ceteras and the and-so-forths who do the work." First Lady Eleanor Roosevelt, Governor Herbert Lehman of New York, and Mayor Fiorello La Guardia all served on the Honorary Committee of the American Common, and all could well have seen the need for—and political value of—promoting America's diversity as a magnet for immigrants. And who better to give the inauguration speech for the new exhibit than America's most famous immigrant?

*Although billed as a tribute that included American Indians, only 4 names of Native Americans were included among the 600—compared to 79 from Germany, 74 from England, and 42 listed as "Negro." There was one name from China.

†Paul Robeson made "Ballad for Americans" (written by Earl Robinson and John LaTouche) an overnight sensation, singing it over CBS radio in November 1939 and the next year to overflow crowds across the country (more than 100,000 estimated in Chicago, and the largest attendance in the history of the Hollywood Bowl). Emphasizing America's multiracial, multiethnic background and the need for equality ("Man in white skin can never be free while his black brother's in slavery"), the words also hail the nation's potential ("Our country's strong, our country's young, and its greatest songs are still unsung") and was considered patriotic enough to be sung, though not by Robeson, at the 1940 National Convention of the Republican party; see Armentrout and Stuckey, *Paul Robeson's Living Legacy,* 11.

But Einstein's speech did not fit into a patriotic scenario. While he acknowledged the "fine and high-minded idea" behind the Wall of Fame exhibit, he pointed out that America "still has a heavy debt to discharge for all the troubles and disabilities it has laid on the Negro's shoulders." Perhaps that's why the speech has never been published in its entirety (for the full text, see Part II, Document B).

> It is a fine and high-minded idea, also in the best sense a proud one, to erect at the World's Fair a wall of fame to immigrants and Negroes of distinction.
>
> The significance of the gesture is this: it says: These, too, belong to us, and we are glad and grateful to acknowledge the debt that the community owes . . . those who are often regarded as step-children of the nation. . . . If, then, I am to speak on the occasion, it can only be to say something in behalf of these step-children. . . .
>
> As for the Negroes, the country has still a heavy debt to discharge for all the troubles and disabilities it has laid on the Negro's shoulders, for all that his fellow-citizens have done and to some extent still are doing to him. To the Negro and his wonderful songs and choirs, we are indebted for the finest contribution in the realm of art which America has so far given to the world. And this great gift we owe, not to those whose names are engraved on this "Wall of Fame," but to the children of the people, blossoming namelessly as the lilies of the field.
>
> In a way, the same is true of the immigrants. They have contributed in their way to the flowering of the community, and their individual striving and suffering have remained unknown.
>
> One more thing I would say with regard to immigration generally: There exists . . . a fatal miscomprehension. Unemployment is *not* decreased by restricting immigration. For [unemployment] depends on faulty distribution of work among those capable of work. Immigration increases consumption as much as it does demand on labor. . . .

The Wall of Fame arose out of a high-minded ideal; it is calculated to stimulate just and magnanimous thoughts and feelings. May it work to that effect.[14]

With reservations, Einstein continued to support Roosevelt, sending congratulations on his reelection in 1940, yet, privately at least, castigating him for failing to commit America more quickly to the anti-Nazi fight. At one point, as German planes were bombing London during the summer of 1940, Einstein called Roosevelt's abstention "disastrous."[15] But after Pearl Harbor, Einstein cheered America's entry into the war he called "a struggle between those who adhere to the principles of slavery and oppression and those who believe in the right of self-determination." In a telephone message to be broadcast to the German people, he said he felt "particularly fortunate to be an American. America is today the hope of all honorable men."[16] J. Edgar Hoover did not include these remarks in Einstein's FBI file.

Robeson, too, became a Roosevelt backer—after the Nazis invaded the Soviet Union in June 1941. Until then, the Nazi-Soviet "nonaggression" pact of 1939 kept Robeson on the sidelines, following the pro-Moscow line of nonintervention. As soon as Hitler's troops crossed into Soviet territory, violating the pact, the Soviets changed their line and so did millions of their supporters around the world, including Robeson. In 1942, he linked America's war effort to his lifelong fight for equal rights: "A victory for Hitler would be the worst thing which could happen to my people. It would mean we would all be consigned to slavery for I don't know how long. Therefore, the salvation of the Negro people lies in the overthrow of fascism."[17]

On the morning of August 16, 1942, when he arrived in Princeton to do *Othello,* Robeson told *The Princetonian* that the racial situation in America "has improved," adding, "If the Administration . . . permits the Negro to help in the war effort as he wants to, I don't think there will be any trouble." The "if" in Robeson's forecast turned out to be the key word. While more than a million black Americans ultimately served in World War II, the vast majority were consigned to "support" units. Only 5 percent saw combat—in all-black units usually commanded by white officers.[18] It was something less than an all-out commitment to equality, as well as a clear signal of struggles to come.

CHAPTER 7

The Home Front

Barred from working on the Manhattan Project and denied security clearance by the army (based on Hoover's memo), Einstein remained in Princeton, wielding a pen as his main weapon to support the war effort. There is no evidence that he knew he had been banned from working on the bomb, but he may well have suspected.[1]

After his backstage visit in 1942, Einstein did not meet Robeson again until after the war. In the meantime, both supported several war-related organizations such as the American-Soviet Friendship Council.[2] Following the momentous Soviet victory at Stalingrad, Einstein wrote a fund-raising letter for the council and sent greetings to several of the group's rallies, where Robeson was a frequent speaker. Like the Russian War Relief meeting mentioned earlier, the council attracted an array of prominent—and politically moderate—supporters. Months after the war's end, one support rally in Madison Square Garden drew thousands of people—and at least one of J. Edgar Hoover's agents, who reported: "Messages of greetings . . . were sent by President Truman, Admiral King, General Eisenhower,* Secretary of War Patterson, Eleanor Roosevelt, Professor ALBERT EINSTEIN and others."[3] (Throughout his FBI dossier, Einstein's name appears in capital letters.)

Would Einstein have agreed to work on the bomb if he had been asked? At least some signs point to "yes." For more than a

*General (and future President) Eisenhower told a House committee in 1945 that "nothing guides Russian policy so much as a desire for friendship with the United States" (Duberman, *Paul Robeson*, 253).

year during the war, Einstein worked as consultant for the U.S. Navy—which gave him security clearance even though the army refused—on problems relating to high explosives; also, in 1943, he donated a hand-rewritten copy of his 1905 special relativity theory to be auctioned off in a fund-raising effort for war bonds.[4]

Staying in Princeton during the war meant, too, that Einstein could maintain his habit of walking around town, including his Witherspoon walks, with stops at Ms. Pannell's house and others', to sit on the porch and chat. Besides the weather, the war, and Robeson's performance in *Othello,* their chats inevitably included events on "the home front," where, in several cities around the country, a very different kind of war was simmering.

Walter Jackson, a thirty-five-year-old African American defense worker, his wife, and five children moved into the new federally sponsored Sojourner Truth Houses in Detroit in the spring of 1942. "We are here now and let the bad luck happen," Jackson said. "I have only got one time to die and I'd just as soon die here." Jackson, a short, wiry 130-pound former UAW-CIO shop steward, had taken an active part in militant auto sit-down strikes in 1937. The Sojourner Truth project had been built to ease Detroit's acute housing shortage—tens of thousands of white and black workers had relocated from the South to work round-the-clock producing tanks and planes. The new housing for blacks was originally planned for a black neighborhood, but the federal government switched and built it in the midst of a white, working-class community. Integrated housing was not permitted by the Federal Public Housing Authority.

On February 27, 1942, with a cross burning in a nearby field, 150 angry whites picketed the project vowing to keep out blacks. By dawn the following day, the crowd had grown to 1,200, many of whom were armed. The first black tenants, rent paid and leases signed, arrived at 9 a.m. but left the area fearing trouble. It wasn't long in coming. Fighting began when two blacks in a car attempted to run through the picket line. Clashes between white and black groups continued into the afternoon when sixteen mounted police attempted to break up the fighting. Tear gas and shotgun shells were flying through the air. Officials announced an indefinite postponement of the move. Finally, at the end of April, despite continuing resentment, black families moved into the project. Walter Jackson and his family were the first to move in. Detroit Mayor Edward Jeffries ordered police and state troops to keep the peace during the move. A score of white women, some pushing baby carriages, waved American flags and paraded briefly along Conley Avenue north of the project. In the following weeks, the Klan burned several crosses, but the United Auto Workers supported the African American tenants, organizing integrated groups of white and black workers for defense squads and picketing, "with much of the leadership coming from left-wing African American workers from Ford's River Rouge UAW Local 600."[5]

Although the initial Sojourner Truth confrontations resulted in no fatalities, it was a warning of what was to come. According to a *Detroit News* report:

> Early in June 1943, 25,000 white workers at a Packard plant that produced engines for bombers and PT boats, stopped work in protest of the promotion of three blacks. A handful of agitators whipped up animosity against the promotions. During the strike a voice outside the plant reportedly shouted, "I'd rather see Hitler and Hirohito win than work beside a nigger on the assembly line." . . . The Ku Klux Klan and the feared Black Legion were highly visible in the plant.[6]

The Packard strike ended only when the federal government threatened to fire the strikers for disrupting the war effort. But street fighting broke out in early August at the Belle Isle amusement park and escalated into a full-scale riot. When soldiers from a nearby army base joined the white mobs, President Roosevelt sent in federal troops to end the fighting—but not before 35 blacks and 9 whites had been killed and police had arrested 1,800 (mostly black) people.

When he was asked to join 137 other "noted Americans" in a statement urging President Roosevelt "to use all wisdom to prevent a repetition of the horrors of Detroit elsewhere in our country," Einstein quickly agreed. The statement was more plea than protest. The NAACP and others assailed Detroit's handling of the riot, charging that police targeted blacks while turning their backs on white atrocities—85 percent of those arrested were black, and all 17 people killed by police during the riot were black.[7] By contrast, the statement Einstein signed urging Roosevelt to prevent a repetition "of the horrors of Detroit" could not have been milder. Nonetheless, it drew the attention of J. Edgar Hoover's loyalty-overseers. HUAC, exhibiting its unique analytical ability, quickly concluded that the Detroit race riots were "Communist-inspired." The FBI made no attempt to investigate the activities of the KKK or the Black Legion in Detroit, or the widespread reports that the two racist groups had whipped up the initial violence that became the Detroit riots of 1943. But Hoover's agents did list the Einstein-signed appeal to Roosevelt as "derogatory information" in the scientist's dossier.[8]

For Einstein, signing the Detroit-riot statement signaled his determination to use his famous name—publicly—on the race issue. His future statements would not be so moderate.

Like Einstein, Robeson walked on two political legs during the war. While heavily involved in supporting the fight against fascism abroad, he continued to speak out against racism at home. In one little-known action, Robeson and a group of eight black newspapermen attended the 1943 annual meeting of Major League Baseball team owners, and met with baseball commissioner Kenesaw Mountain Landis to demand baseball's integration. After Robeson addressed the owners for twenty minutes, they applauded—but did nothing. This was three years before Jackie Robinson signed with the Brooklyn Dodgers.

One evening in 1942, Einstein's walk through the Witherspoon community took him not to the hospital but to the Witherspoon School for a meeting of the Princeton NAACP. The school auditorium was crowded, NAACP secretary Walter White had come from New York, and signs on the wall urged members to "Buy War Bonds." Einstein not only joined the organization, but he later also helped sponsor the NAACP's new Legal Defense Fund. He publicly endorsed the campaign to stop the extradition from New Jersey of Sam Buckhannon, who had escaped from a Georgia chain gang after serving eighteen years for stealing a pack of cigarettes. The protest movement eventually succeeded in blocking Buckhannon's extradition.

Fanny Floyd grew up around the Princeton NAACP. "My parents were active, especially my father, so I went to a lot of meetings. Princeton had a very strong NAACP organization and national leaders would come down sometimes." But one meeting left a special impression. In a round-table discussion with other African American Princetonians she recalled:

> Yes, that meeting where Einstein came was held in the Witherspoon School—around 1942—I think it was a bond rally, you know, raising money to buy bonds to support the Second World War. I was a teenager but everyone knew Einstein. We had all seen him on the street, walking on Nassau Street and down Witherspoon.
>
> I didn't actually talk to Einstein at that meeting. It was a big meeting. I think Walter White, the national head of the NAACP, was there—I remember somebody came in a limousine. (Actually, I kind of think it was [Congressman] Adam Clayton Powell in that limo.) Anyway, there were a lot of folks around Einstein. And I guess at my age, I was in high school, I was probably sitting in the back.

Was it unusual for a teenager to attend NAACP meetings? "I didn't think it was strange to go to so many NAACP and other meetings—don't forget it wasn't so long ago that the KKK was still active in the area—they had burned a cross in Hightstown not too long ago."

The discussion turns to more recent racism: "They burned a cross in *Princeton* in the 1980s," Henry Pannell says.*

"And they're still out there," adds Penney Edwards-Carter.

"Even today," says Shirley Satterfield, "the town still has that master-slave complex. It's still there but it's more subtle—it's not as clear as when we were growing up, but there's still that aura of segregation here, there are still certain neighborhoods—the new rich."

Penney Edwards-Carter agrees: "I've been walking down the street and have women put their pocketbooks on the other side of their shoulder and I'm thinking to myself how many people have I hit in the head lately? I think people just see a black face, they don't see the person—we're invisible. And the media plays it up. I think whites who don't really deal with a whole lot of black people think we're a bunch of criminals."

Hoover's FBI recorded Einstein's support for the NAACP and defense of Sam Buckhannon as more "derogatory information" in Einstein's file.[9] "Derogatory" was not a term the FBI chief tossed about casually. If a report wasn't what he considered "derogatory," Hoover wasn't interested. A revealing handwritten notation on an internal FBI memo in Einstein's file declares: "No dissemination, as report contains no pertinent derogatory information."[10]

It's difficult to imagine a motive for Hoover's decision to label Einstein's NAACP activity in Princeton and his mild statement on Detroit as "derogatory"—except for racism. Since his death in 1972, Hoover's public standing has plummeted. At least part of the reason was the exposure of his "dirty tricks" campaign against Martin Luther King Jr. But the FBI chief's—and the FBI's—racism began long before King came on the scene.

*In 1971, the KKK burned a seven-by-four-foot cross at the corner of Route 206 and Princeton Pike as part of a recruitment drive; see Kathryn Watterson, *I Hear My People Singing*, a soon-to-be-published oral history of Princeton's African American community.

Hoover learned his racism as the fourth "R" in the white-only schools he attended, along with white-only churches in the white-only world that was his sector of one of America's most segregated southern cities, Washington, D.C. One of Hoover's first Justice Department security files, set up in 1919, was labeled "Negro Activities." It denounced back-to-Africa advocate Marcus Garvey as "a notorious negro [*sic*] agitator" and claimed that Garvey's newspaper, *Negro World,* "upheld . . . Soviet Russian rule" and engaged in "open advocation of Bolshevism." The same files attacked other black-owned newspapers, including A. Philip Randolph's *Messenger* and *The Crisis,* edited by W.E.B. Du Bois: "[S]omething must be done to the editors of these publications as they are beyond doubt exciting the negro [*sic*] elements of this country to riot and to the committing of outrages of all sorts."[11] There is no evidence of any Hoover file on the Ku Klux Klan, which at that time was on a rampage of lynching and cross-burning.*

The FBI chief "tried to avoid civil rights cases whenever he could," according to Hoover biographer Richard Gid Powers, a political moderate. "By the fifties, this had become fixed Bureau policy. . . . Hoover, influenced by his birth in Washington . . . identified instinctively with the racial elite of this country." When four young African American girls were killed in the 1963 terrorist bombing of a Birmingham church, "Hoover quashed

*Some wags might call Hoover an equal-opportunity racist. Like most racism, Hoover's was not restricted to one target group. His anti-Semitism has been widely reported (Jerome, *The Einstein File,* 44); and in a recent biography, *Puppetmaster,* Richard Hack reports that the FBI chief "thought little of the British and the French, and was openly antagonistic to Scandinavians, whom he considered as haughty and pompous" (213). One instance of his anti-Hispanic bigotry came after the attempted assassination of President Truman by Puerto Rican nationalists in 1950, when Hoover said "jokingly": "You never have to worry about a President being shot by a Puerto Rican or a Mexican. They don't shoot very straight. But if they come after you with a knife, beware"; see *Washington Post* obituary, May 5, 1972.

an FBI probe into the . . . bombing despite strong evidence against Robert Chambliss and fellow Ku Klux Klan members."[12]

Inside the FBI, for decades virtually the only African American agents were Hoover's chauffeur, "Special Agent" James Crawford; two other chauffeurs in La Jolla and Miami (Hoover vacationed there); a messenger, Worthington Smith; and Hoover's in-office coat-taker, Sam Noisette. When World War II started, Hoover gave all of them the title of FBI agent so they would not be drafted.[13]

The "Good War" ended with the dropping of two American atom bombs on Hiroshima and Nagasaki—and the death of some two hundred thousand Japanese civilians—on August 6 and 9, 1945. If America was united behind the antifascist war, the atom bombs that ended the war also exploded that unity. A growing number of critics, many from religious groups, assailed "the indiscriminate slaughter of noncombatants" as "ghastly" and "morally indefensible." Others, including scientists who had worked on the bomb, worried about how the United States might act as the world's only nuclear power. One group of these Manhattan Project alumni formed the Emergency Committee of Atomic Scientists and enlisted Einstein (even though he had been banned from the project itself) as their chairman. Polls showed that while most Americans supported the bombing, only 47 percent felt it had reduced chances for another war and nearly two-thirds felt there was "a real danger" that atom bombs would some day be dropped on America. On CBS radio, Edward R. Murrow reported an unprecedented national "sense of uncertainty and fear . . . a realization that survival is not assured."[14] A year after the bombing, John Hersey's *Hiroshima* quickly became an overnight best-seller: "Not many people walked in the streets, but a great many sat and lay on the pavement, vomited, waited for death, and died. . . . [The bomb] killed ninety-five percent of the people within half a mile of the center, and many thousands who were farther away."

The atom-bombing of hundreds of thousands of nonwhite civilians, in addition to widespread anti-Japanese cartoons and editorials in the U.S. press, and politicians' racially charged remarks,* aggravated the racial fault line that had long divided the country. While 98 percent of the white press supported the nuclear attack on Japan, the black press was more divided and, within a few months of the war's end, increasingly critical.[15] By mid-September, the *Chicago Defender* had carried articles by poet Langston Hughes, scholar W.E.B. Du Bois, and NAACP secretary Walter White, all siding more with the Japanese victims than with the U.S. action. An editorial in the NAACP's *Crisis* said the bombing of Hiroshima had raised the question, "Who is barbarian and who is civilized?" Noting that "the 7,000 black workers at [the Manhattan Project's research center in] Oak Ridge lived in segregated, inferior housing and performed menial jobs only, with no school provided for their children [unlike white workers]," the *Washington Afro-American* asserted the United States had spared the Germans who, "after all, represent the white race," and "saved our most devastating weapon for the hated yellow men of the Pacific." The paper concluded that Hiroshima had "revived feelings in some quarters that maybe the Allies are fighting a racial war after all."[16]

By the end of 1945, it was clear that the "racial war," if it was one, may have ended in the Pacific, but not on the home front.

*Two weeks after he ordered the atom bombs dropped, President Harry S. Truman, in response to critics, declared: "When you deal with a beast, you have to treat him as a beast"; see *New York Times*, August 10, 1945, 6.

CHAPTER 8

Civil Rights Activist

It was in 1946 that Einstein became what today we might call a civil rights activist, and again linked arms with Paul Robeson. More than a million African American soldiers were coming home—some would never come home—from the war against fascism, and many black GIs had adopted a "Double-V" watchword, for victory over Nazism abroad and victory over racism at home. "Many of us had illusions that, as a result of the war, the whole system of second-class citizenship and discrimination against blacks would be ended," one black GI later wrote, explaining:

> I was in the 92nd Division, along with ten or twelve thousand other black troops. In Europe, the men of the 92nd were regarded as heroes. We liberated a number of Italian towns, including Lucca and Pisa, and when the Italian people saw these brown troops coming into their community they just hailed us as conquering heroes. So when we came back to the United States, we expected to be treated as if we had made a contribution; we didn't like coming back into a Jim Crow scene. Most of the enlisted men in the 92nd were from the South, and it was ironic for them to return to a country for which they risked their lives, and they still had to go to the back of the bus, could not sit downstairs in the movie theater, and could not leave the plantation except with a pass from the owner.[1]

But the defeat of the Aryan super-race in Berlin seemed to have no effect on the Aryan super-racists in Mississippi,

Alabama, Georgia, and the like, except to intensify their terrorism. Mississippi's senator Theodore Bilbo, an admitted member of the Ku Klux Klan, called for white men "to employ any means" to stop blacks from voting.[2] In the first year after the victory over fascism, a wave of antiblack violence, mostly but not only in southern states, killed fifty-six African Americans. Some were outright lynchings, others involved police shootings of unarmed victims, with returning veterans the most frequent targets.[3]

Einstein's 1946 civil rights activism began with the publication in *Pageant* magazine of his article "The Negro Question," arguably his most eloquent challenge to racism in America. (For complete text, see Part II, Document C.) Writing "seriously and warningly," Einstein declared:

> There is . . . a somber point in the social outlook of Americans. Their sense of equality and human dignity is mainly limited to men of white skins. Even among these there are prejudices, of which I as a Jew am clearly conscious; but they are not important in comparison with the attitude of the "Whites" toward their fellow-citizens of darker complexion, particularly toward Negroes. The more I feel an American, the more this situation pains me. I can escape the feeling of complicity in it only by speaking out.
>
> Many a sincere person will answer me: "Our attitude toward Negroes is the result of unfavorable experiences which we have had by living side by side with Negroes in this country. They are not our equals in intelligence, sense of responsibility, reliability."
>
> I am firmly convinced that whoever believes this suffers from a fatal misconception. Your ancestors dragged these black people from their homes by force; and in the white man's quest for wealth and an easy life they have been ruthlessly suppressed and exploited, degraded into slavery. The modern prejudice against Negroes is the result of the desire to maintain this unworthy condition. . . .

> I believe that whoever tries to think things through honestly will soon recognize how unworthy and even fatal is the traditional bias against Negroes.

Finally, Einstein urged the mostly white *Pageant* readership to get involved in the civil rights struggle:

> I do not believe there is a way in which this deeply entrenched evil can be quickly healed. But until this goal is reached there is no greater satisfaction for a just and well-meaning person than the knowledge that he has devoted his best energies to the service of a good cause.

In February, a group of black veterans in Columbia, Tennessee, fired their guns at a white lynch mob attacking their community ("Mink Slide"), and wounded two of the (armed) whites. When he heard that armed African Americans had shot at whites, Tennessee's governor immediately sent in 500 state troopers who roped off the community and, firing submachine guns, destroyed virtually every black-owned business in the four-square-block area. The troopers arrested more than one hundred black men, and police later shot and killed two of the detainees inside the jail during what Tennessee officials called "a spontaneous outburst." Twenty-five others were indicted for "attempted murder."[4]

Reading about the case in the *New York Times,* Einstein could not have missed the headline, NAACP TELLS TRUMAN SHOOTING BY TENNESSEE TROOPERS WAS WORTHY OF NAZIS, and the statement of a young NAACP lawyer named Thurgood Marshall, chief defense attorney for the twenty-five: "The actions of the Tennessee State troopers in roping off the Negro section of Columbia, Tenn. and firing at will and indiscriminately was closer to . . . German storm troopers than any recent police action in this country."[5] That same month, a leading Democratic party official warned that "the revival of the Ku Klux Klan" was spreading through Indiana, Illinois, and Ohio, where they "were trying to capitalize on the discontent of racial groups of German origin."[6]

With his indelible memories of Nazism, Einstein could not have missed the echoes. Shortly after Thurgood Marshall's "storm troopers" statement, Einstein publicly joined the National Committee for Justice in Columbia, Tennessee, headed by Eleanor Roosevelt.[7] A few weeks later, he traveled to Lincoln University in Pennsylvania to accept an honorary degree and speak to the students.

Einstein almost never spoke at universities during the last twenty years of his life. His increasingly frail health made travel difficult, but mainly he considered the pomp and ceremony of degree presentation to be "ostentatious."[8] Some may find it remarkable that Einstein chose to break his no-college rule by going not to an Ivy League producer of prestigious degrees,* but to a traditionally black university. (Chartered in 1854, Lincoln was "the first institution anywhere in the world to provide a higher education in the arts and sciences for male youth of African descent.")[9] But for Einstein, the sixty-mile trip from Princeton to Lincoln was not a casual choice. His visit was "in a worthwhile cause," he told the assembled students and faculty. "The separation of the races [segregation]," he declared, "is not a disease of colored people, but a disease of white people," adding, "I do not intend to be quiet about it."[10]

Disease? To appreciate today Einstein's choice of the word requires examining specific symptoms of the segregation sickness so widespread in America eighty years after the Civil War. Black soldiers, as we have noted, when allowed into combat at all, fought in segregated units under white officers. To a true believer, segregation *always* came first, even before patriotism: vowing never to fight "with a Negro by my side," a twenty-eight-year-old West Virginia lawyer—and future senator—named

*In 1936, Einstein refused to take part in Harvard's Tercentenary ceremony because German universities were participating

Robert Byrd wrote to Mississippi's Senator Bilbo in 1945: "Rather I should die a thousand times, and see Old Glory trampled in the dirt never to rise again, than to see this beloved land of ours become degraded by race mongrels, a throwback to the blackest specimen from the wilds." (Byrd has since undergone a sea-change in his views on race.)[11]

Racial segregation was the rule in most of America in May 1946, with separate and unequal public and private facilities from housing and schools to buses and beaches throughout the south and many other parts of the country, including, as we've seen, Princeton, New Jersey. Some textbooks and even some documentary films have pictured the separate (and decidedly unequal) waiting rooms in southern bus and train stations, and even the separate drinking fountains marked "colored" and "white." But the disease went deeper.

Even the blood donated to save lives was donated at racially segregated blood banks (when blacks were allowed to donate at all), with "white" and "colored" blood kept in separately labeled storage units. In 1942, in the midst of a world war, the American Red Cross met in Washington and concluded that, while there is no difference in the blood of the races, "most men of the white race objected to blood of Negroes injected into their veins." No one apparently asked, according to one writer, "how many white soldiers, hemorrhaging from a gaping wound on the battlefield and sinking into a coma, would stop a medic from giving them the 'wrong color' plasma." The policy of racially segregating blood continued in some parts of this country well into the 1960s.[12]

The Lincoln students in Einstein's audience, of course, knew most of this in 1946. "On Friday, May 3rd, a very simple man came to Lincoln University," one student wrote a few days later in the school newspaper, *The Lincolnian*:

> His emaciated face and simplicity made him appear as a biblical character. Quietly he stood with an expression of questioning wonder upon his face as . . . President Horace

Mann Bond conferred a degree. Then this man with the long hair and deep eyes spoke into a microphone of the disease that humanity had. In the deep accents of his native Germany he said he could not be silent. And then he finished and the room was still. Later he lectured on the theory of relativity to the Lincoln students.

That night, Albert Einstein went back to Princeton.[13]

Before returning home, Einstein had dinner at the home of Professor Laurence Foster and his family. Foster's daughter Yvonne recalls, "The faculty had been warned that Einstein was very shy and low key, and, in fact, he was quiet and spoke very little at dinner, but he became amused by [her younger brother] Larry's Pennsylvania Dutch accent and couldn't help but smile during the conversation." She adds, "We were honored" that during the degree-presentation ceremony, "Einstein wore Professor Phillip Miller's academic robe and daddy's mortarboard."

"I was very happy to know that my boy had an opportunity to see Dr. Einstein," one student's mother told Lincoln's President Bond shortly after the event. In a letter to Einstein, Bond relayed the mother's words, adding his own thanks: "All of us are as grateful as this humble mother."[14]

Einstein's choice of Lincoln, as well as his words, clearly seemed intended to send a message to a wider audience. But the media then—like the media now—had different news priorities. While almost all of Einstein's public speeches and interviews were extensively reported by major newspapers[15]—even sticking out his tongue at a reporter made the front pages—the mainstream media treated the address by the world's most famous scientist at the nation's oldest black university as a nonevent. Only the black press gave Einstein's speech meaningful coverage. The *Philadelphia Tribune* and *Baltimore Afro-American* carried firsthand front-page reports with photos of Einstein receiving the honorary degree from Lincoln's president Horace Mann Bond and lecturing on relativity to Lincoln students. The *Tribune*'s headline, stretching across half the front page, read:

EINSTEIN AWARDED PORTFOLIO
In Historic Campus Ceremony

Other black papers covering the story included the *New York Age*, *New York Amsterdam News* ("Einstein: Race Problem a Disease of 'White Folks'"), and *Pittsburgh Courier*.[16] All included photos.

No one has yet found a copy or transcript, or even notes, of Einstein's speech at Lincoln, nor has it ever been quoted in the plethora of Einstein biographies and anthologies. What follows are excerpts from his ten-minute speech based on the report in the *Baltimore Afro-American* of May 11, 1946. The article, by J. W. Woods, is datelined LINCOLN UNIVERSITY, Pa.:

> The only possibility of preventing war is to prevent the possibility of war. International peace can be achieved only if every individual uses all of his power to exert pressure on the United States to see that it takes the leading part in world government.
>
> The United Nations has no power to prevent war, but it can try to avoid another war. The U.N. will be effective only if no one neglects his duty in his private environment. If he does [neglect it], he is responsible for the death of our children in a future war.
>
> My trip to this institution was in behalf of a worthwhile cause. There is separation of colored people from white people in the United States. That separation [segregation] is not a disease of colored people. It is a disease of white people. I do not intend to be quiet about it.
>
> The situation of mankind today is like that of a little child who has a sharp knife and plays with it. There is no effective defense against the atomic bomb. . . . It can not only destroy a city but it can destroy the very earth on which that city stood.

The *New York Times* of May 4, 1946, carried a brief item on page 7, with a total of one sentence about the speech: "Dr. Einstein said he believed there was 'a great future' for the Negro [and] asked the students 'to work long and hard and with lasting patience.'" Assuming this sentence was taken from the same speech (none of the reports in the black press cite or allude to that sentence or anything similar to it), it is interesting to contrast what the *Times* singled out to publish with what the black press reported.

Einstein's equal-rights advocacy was not limited to articles and speeches. At the end of the summer, Paul Robeson asked Einstein to join him as co-chair of a new group, the American Crusade to End Lynching (ACEL), which planned to hold a protest rally on September 23—the anniversary of the Emancipation Proclamation—in Washington. Its purpose would be to demand the passage of a federal antilynching law. Einstein agreed.

If Einstein had not read about the wave of lynchings that summer—they were rarely mentioned in the major media—he might well have heard about them from conversations with Ms. Pannell and others during his frequent visits to the Witherspoon community. There, as in African American neighborhoods around the country, news of another unpunished lynching seemed to arrive almost every week, often reported only in the black press.[17] That changed dramatically on July 27: millions of Americans were shocked when the *New York Times* reported a grisly lynching in Monroe, Georgia, on its front page:

GEORGIA MOB OF 20 MEN MASSACRES
2 NEGROES, WIVES; ONE WAS EX-GI

It is impossible to feel the impact of the story without reading at least some of it:

> MONROE, Ga., July 26—Two young Negroes, one a veteran just returned from the war, and their wives were lined up last night near a secluded road and shot dead by an unmasked band of twenty white men. . . .
>
> The ghastly details of the multiple lynching were told today by Loy Harrison, a well-to-do white farmer who was tak-

> ing the Negroes to [work on] his farm and was held at gunpoint [when his car was waylaid] by the mob. . . .
>
> The Negro men were taken out of the car first and led down a side road. The women were held at the automobile. Then a member of the mob said that one of the women had recognized him.
>
> "Get those damned women, too," the mob leader shouted. Several of the men then . . . dragged the shrieking women from the automobile.

The *Times* account also said that the victims had been shot at least sixty times, and the bodies were "scarcely recognizable" because of the mass of bullet holes. Two days later, the *Times* reported:

> MONROE, Ga., July 28—Close relatives of two of the four Negroes killed by a white mob here Thursday failed to appear at funeral services today and friends voiced the belief that they were "too frightened" to appear.

What had been going unreported and unpunished for months—lynching in America—was suddenly an international issue, and the public attention helped spur plans for the ACEL's Washington demonstration. Some three thousand antilynching protesters turned out on September 23. Einstein was too ill to attend, but he gave Robeson a letter for President Harry Truman:

> May I wholeheartedly endorse the aims of this delegation, in the conviction that the overwhelming majority of the American people is demanding that every citizen be guaranteed protection against acts of violence. To insure such protection is one of the most urgent tasks for our generation. (See full text in Part II, Document E.)

When a multiracial ACEL delegation, led by Robeson, met with Truman in the White House after the protest rally, the president told them the time was "not right" for a federal antilynching law.[18] One account of the confrontation that followed reported that Robeson told Truman "returning [black] veterans are . . . determined to get the justice here they have fought for abroad . . . and their restiveness might produce an emergency situation."[19]

The FBI's dossier on Einstein says virtually nothing about his *Pageant* article or his talk at Lincoln. (The dossier in fact has very little about anything Einstein wrote or said—it reports mainly on his organizational affiliations.) But when it came to the American Crusade to End Lynching, Hoover's informers kept busy. Einstein's dossier has no fewer than twelve pages of "items" on the ACEL (see Part II), including "this Crusade has all the earmarks of another Communist attempt to instill racial agitation."[20] But the nation's chief federal law enforcement officer, J. Edgar Hoover, did nothing about the lynching—just as he ignored the Ku Klux Klan for decades. "Despite the KKK's hand in murder, lynching and assault," the conservative *Life* magazine reported years later, "it was not put under FBI surveillance until 1964."[21]

Perhaps it is not completely accurate to say that Hoover did nothing about lynching in 1946. On September 17, he sent a memo to the attorney general, asserting, "It is a mistake for the [Justice] Department and the Bureau [FBI] to enter these mob-violence cases unless and until there is some showing of a federal violation."[22]

The Crusade did not end lynching nor get an antilynching law enacted, despite the public efforts of two of America's—and the world's—most popular figures at the time,* as well as support from other prominent citizens, both white and black.[23] But their protest did bring the debate to a wider audience—one more ripple in the current that would swell into a giant wave of people-

*A 1947 Gallup Poll found Robeson was one of the America's "favorite people," finishing as one of forty-eight runners-up to the top ten (*Look* magazine, June 24, 1947). David Levering Lewis (*W.E.B. Du Bois,* 515) calls Robeson "the most famous black man in America and one of the few Americans of any race or creed who was so well known that he could have dropped into a remote village in Uzbekistan or India and been recognized."

protests in the 1960s civil rights movement. For Einstein and Robeson, the ACEL was their first shared antiracist organization. There would be more.

Einstein's sharpest attack on racism came in a message he sent in September to a conference of the National Urban League.[24] (See the complete text in Part II, Document F.) "The worst disease under which the society of our nation suffers," he declared, "is . . . the treatment of the Negro." He added that racism is "the scorn of the principle of the Fathers who founded the United States that 'all men are created equal.'"

> One can hardly believe that a reasonable man can cling so tenaciously to such prejudice, and there is sure to come a time in which school-children in their history lessons will laugh about the fact that something like this did once exist.
>
> Today, however, this prejudice is still alive and powerful. The fight against it is difficult as is the fight on every issue in which the enemy has thoughtlessness and a fatal tradition on its side. . . .
>
> First we must make every effort [to insure] that the past injustice, violence and economic discrimination will be made known to the people; the taboo, the "let's-not-talk-about-it" must be broken. It must be pointed out time and again that the exclusion of a large part of the colored population from active civil rights by the common practices is a slap in the face of the Constitution of the nation.
>
> We must strive [to ensure] that minorities be protected against economic and political discrimination as well as against attack by libelous writings and against the poisoning of youth in the schools. These endeavors are important, but not as important as the intellectual and moral enlightenment of the people. For example, it is clear that in the South, through economic pressure on the colored population, wages in general are kept down, and that thereby the majority of the people [white and black] suffers pauperization through reduced buying power. If the majority knew of the root of this evil, then the road to its cure would not be long. . . .

There is a thing which one could call the moral climate of a society. In one society, there may be a preponderance of distrust, malice and ruthless egotism; in another, the enjoyment of beauty and of the blooming life, compassion for the suffering of one's fellow-men and rejoicing in their happiness. . . .

One thing is certain: No mechanism can give us a good moral climate as long as we have not freed ourselves from the prejudices to whose defeat you are devoting yourselves.

Before 1946 was over, Einstein signed on with the National Committee to Oust Bilbo, joining a bevy of entertainment stars, including Gene Kelly, Leonard Bernstein, Fannie Hurst, Oscar Hammerstein II, David O. Selznick, Katherine Dunham, and John Garfield. Other sponsors included Adam Clayton Powell Jr., Alain Locke, and co-chairmen Quentin Reynolds and Vincent Sheehan. The committee distributed posters, pamphlets, buttons, and 185,000 Oust-Bilbo petitions in thirty-two states. The group also publicized Bilbo's corruption and his role in organizing violent attacks on African Americans attempting to vote in Mississippi.[25]

The U.S. Senate has never been exactly a hub of diversity—three African American senators (one at a time) since Reconstruction. But in the twentieth century, no single senator identified himself with white supremacy as publicly and unashamedly as Mississippi's senator Theodore Bilbo. Up for reelection in November 1946, Bilbo was interviewed in August on the popular radio program *Meet the Press*. Host Lawrence Spivak asked him about the KKK:

Spivak: Do you think you would get any Klan support now?
Senator Bilbo: I do.
Spivak: You never left the Klan, in effect?
Bilbo: No man can leave the Klan. He takes an oath . . . once a Ku Klux, always a Ku Klux.[26]

Three months later, Bilbo was reelected to his third Senate term. In December, Einstein's FBI file (Section 8) tells us:

> [Informant's name blacked out] furnished the Bureau with literature distributed by the National Committee to Oust Bilbo, sponsored by the Civil Rights Congress. . . . Included in this material was a letter dated Dec. 4, 1946, signed by Quentin Reynolds and Vincent Sheehan. . . . The names of 55 members of this Committee were set out including that of ALBERT EINSTEIN.

The Oust Bilbo Committee even produced a song it distributed in sheets and on records through its branches in thirty-three states. Perhaps it was influenced by all the creative talent drawn to the committee, or perhaps by the tradition, still very much alive in 1946, of Woody Guthrie, the working people's troubadour. Called "Listen Mr. Bilbo,"* the song had a broad, populist appeal, emphasizing America's diversity and Bilbo's anti-everybody-ism:

> You don't like Negroes, you don't like Jews,
> If there's anyone you do like, it sure is news.[27]

The FBI pursued the Committee to Oust Bilbo partly because it was antiracist, a trait Hoover considered dangerous and, often, "subversive." But the Bureau targeted the committee also because it was affiliated with the Civil Rights Congress, a group that publicly defended the Communist party.[28] The CRC counted Paul Robeson among its leaders and, among its endorsers, Albert Einstein.

*In 1945, New Yorkers Bob and Adrienne Claiborne wrote the lyrics to "Listen Mr. Bilbo," and set them to a traditional tune, pointing out that "we are all immigrants in this continent." The writer of the insert accompanying the Smithsonian Folkways CD that includes their song comments, "Of course, in making their valid point, the Claibornes didn't address how some of the European arrivals treated those whose ancestors had come here many centuries before. Still, the easily singable song serves as a warning against demagogues."

CHAPTER 9

From World War to Cold War

America emerged on the winning side of two world wars in less than thirty years— having played a key role in both victories—as the only major power that remained relatively unscathed. Japan and Germany had been crushed; Britain badly bombed, and France invaded and occupied. The Soviet Union, bombed, invaded, and partially occupied, lost at least 20 million citizens during World War II. By contrast, "the war rejuvenated American capitalism." U.S. Gross National Product (GNP) jumped from $91 billion in 1939 to $210 billion in 1944, and even more dramatically, in the same years, corporate profits leapt from $6.3 billion to $23.8 billion. With 5.7 percent of the world's population, the U.S. economy accounted for half the world's manufacturing and more than 40 percent of its income.[1]

"It seemed so naturally America's moment," as Peter Jennings and Todd Brewster put it. "No problem was too big for American energy and brainpower . . . the world's largest telescope [was built] in Mount Palomar, California . . . the first civilian forms of penicillin and other antibiotics . . . seemed to promise a cure for all infectious diseases . . . scientists at Bell Laboratory discovered the transistor, and others at the University of Pennsylvania assembled the world's first electronic computer."[2]

William Levitt sold his prefabricated houses by the tens of thousands to returning GI's who, with their growing families, populated Levittowns and dozens of similar tract-housing villages in newly created suburbs across the country. Even as Einstein, John Hersey, and a few others appealed to American consciences—How can you remain silent about the arms race and social injustice threatening the world's future?—millions of

Americans turned inward, away from the world, focusing instead on settling the suburban frontier. The baby boom and the nuclear family were more immediate—and less frightening—than the bomb.

But not all Americans shared in "America's moment"—or in Levitt's towns. From the first Levittown on New York's Long Island, hope-filled families from city apartments resettled into boxy suburban houses with green lawns and white neighbors. Only white. If you were African American, Levittown and its imitators were off limits. In just six years, (mostly suburban) housing starts in America jumped from 114,000 in 1944 to a staggering 1.7 million in 1950,[3] and the mass migration to the suburbs became known as "white flight."

"The biggest gains [from World War II] were in corporate profits," according to the historian Howard Zinn, "But enough went to workers and farmers to make them feel the system was doing well for them."[4] Nonetheless, postwar unemployment continued, especially among African American and women workers, generally the first to be laid off when war production stopped. Black workers "were laid off out of seniority and found to be 'incompetent' to work at jobs they had successfully performed throughout the war." Yet U.S. labor leaders made "little attempt" to reach out to African American workers, women workers, unemployed veterans, or newly arriving Latino workers—"with the exception of a small number of left-wing unions" with interracial membership and leadership.[5]

Despite the gains unions won through a wave of militant strikes in 1946, U.S. labor leaders failed in their campaign—"Operation Dixie"—to organize the South,* leaving millions of

*The exceptions were the few militant, interracial southern unions—such as the Mine Mill and Smelter Workers in Birmingham, the Food and Tobacco Workers in North Carolina, and the packinghouse and maritime workers unions—almost all of which had been organized in the 1930s by the communists.

While anything containing the word "communist" is obviously contro-

low-wage white and black workers out of the prosperity picture and, perhaps just as important, an escape hatch for American companies seeking lower "costs." In the coming years, hundreds of factories and offices would move south for the area's nonunion, low-wage workforce (only to move again later to other countries where they could pay even less). And while southern whites earned little, southern blacks earned even less.

We don't know if Einstein had read about "Operation Dixie" when, in September 1946, he pointed out in his message to the Urban League (cited earlier): "In the South, through economic pressure on the colored population, wages in general are kept down, and . . . thereby the majority of the people [white and black] suffers pauperization through reduced buying power. If the majority knew of the root of this evil, then the road to its cure would not be long." But it is at least an interesting coincidence that Einstein's statement came just as the CIO's southern organizing campaign was getting under way.

Coming at the start of the great "white flight" to "safe" but, as later experience would demonstrate, largely sterile suburbia, another part of Einstein's message to the Urban League seems prescient:

---

versial—Goldfield calls it "a highly contradictory party, difficult . . . to place in a balanced perspective"—the party's role in building virtually the only truly interracial labor unions has been widely recognized: "The communists extended their activities to the organization of black sharecroppers in the heart of the black belt South. They successfully organized large concentrations of African-American workers. . . . It was almost a *sine qua non* during the 1930s: where militant, interracial unionism with strong stances and willingness to struggle for the equality of black workers existed, one would invariably find the CP . . . including such places as the Birmingham steel mills [and beyond the South] the Briggs automobile plants in Detroit, and the Ford River Rouge plant, then the largest manufacturing plant in the United States"; see Goldfield, *The Color of Politics*, 193; Schrecker, *Many Are the Crimes*, 390; also see extensive references in the books by Hudson, Keenan, Kelley, Meier and Rudwick (*Black Detroit*), Naison, and Painter listed in the Bibliography.

> We have become accustomed to judge everything according to tangible and measurable view-points. Food in calories, national and private income in dollars, ownership or non-ownership of a bathtub and of a water-closet. . . . Yet there are other things which are just as decisive for the happiness or unhappiness of the human being. A man may own everything that puts him on a high level according to a crudely materialistic point of view, and have a miserable existence. He may be persecuted by fear, hatred or envy, he may be deaf to merry songs and blind to blooming life. It is similar for social groups, even for whole nations. . . .
>
> One thing is certain: No mechanism can give us a good moral climate as long as we have not freed ourselves from the prejudices to whose defeat you are devoting yourselves.

On the international scene, too, there were flies in America's ointment. If U.S. postwar foreign policy were a slogan, it might have been, "We're Number One!" Indeed, Henry Luce came close to that when the conservative publisher of *Time* and *Life* magazines proclaimed the arrival of "The American Century."

But as policy positions go, catchy slogans leave unanswered questions. In this case, for example: How should the planet's single superpower—economically, technologically and, with a nuclear exclamation point, militarily—relate to the rest of the world? How does the biggest kid on the block treat smaller, weaker kids?

Einstein and Niels Bohr, the renowned Danish physicist, were among those who urged a policy of cooperation and sharing of nuclear information with the Russians. It would not take long anyway, they pointed out, before Soviet scientists developed their own bomb.

Even before the war ended, Bohr, who was at Los Alamos in 1944 (after he and his son had made a dramatic escape in a small boat from Nazi-occupied Denmark), met with Prime Minister

Churchill and President Roosevelt, trying, but failing, to convince the British and American leaders to invite the Soviets to join in planning for cooperative postwar control of nuclear energy. Churchill and Roosevelt reacted to Bohr's peace plan with alarm and suspicion. They warned the Danish physicist that his proposal threatened U.S. and British security, and they ordered him monitored to "make sure he is responsible for no leakage of information, especially to the Russians." Churchill went further, claiming that Bohr was "very near the edge of mortal crimes."[6] The British prime minister and the American president, it seems, had already identified their next enemy.

Einstein's efforts to move Washington toward international cooperation got no further than Bohr's. In 1947, Einstein wrote to Secretary of State George C. Marshall, seeking a meeting. Einstein hoped to convince Marshall that only a world government with military power could prevent a catastrophic nuclear war. Marshall assigned Einstein's letter to Lewis Strauss, then a member of the new Atomic Energy Commission (and the man who later led the effort to discredit J. Robert Oppenheimer). Strauss sent his aide, investment banker William T. Golden, to interview Einstein in Princeton. The scientist criticized increasing U.S. militarism and warned Golden that if Washington tried to impose a *Pax Americana*, "History shows this to be impossible and a certain precursor of war and grief." But the thirty-seven-year-old investment banker dismissed Einstein's views as "naïve in the field of international politics and mass human relations."[7]

If Washington would not be moved, neither would Moscow. Four leading Soviet scientists publicly denounced Einstein's support for world government, charging, among other things, "such ideas represent the imperialist aims of capitalist monopolists."[8]

But Einstein's worries about America's foreign affairs went beyond nuclear weapons. Only four months after the war's end, he told a Nobel anniversary dinner in New York:

> The peoples of the world were promised freedom from fear; but . . . fear among nations has increased enormously since the end of the war. The world was promised freedom from

> want; but vast areas of the world face starvation, while elsewhere people live in abundance. The nations of the world were promised liberty and justice; but even now we are witnessing the sad spectacle of armies . . . firing on peoples who demand political independence and social equality.[9]

Those words could have been a manifesto for the many national liberation movements in former colonies, movements that had gained strength and popular support during their wartime antifascist resistance. In Asia, the world's most peopled continent, revolutionary struggles were ablaze—in Indochina against the French, in Indonesia against the Dutch, in Malaysia against the British, and in the Philippines, where armed rebellion broke out against U.S. troops. And communist guerrillas in China, supported by millions of peasants, were challenging the Western-backed dictatorship of Chiang Kai Shek.* In Africa, too, discontent was on the rise, with militant strikes by workers in French West Africa, Kenya, and South Africa, where 100,000 gold miners stopped work until they were forced back by the army.

Leaving no doubt about his support for national liberation movements, Einstein sent a letter to the Council on African Affairs—Paul Robeson and W.E.B. Du Bois were chairman and vice chairman, respectively—declaring, "No reliable or lasting peace will be possible without the political and economic emancipation of the now subdued and exploited African and colonial peoples." That emancipation, he added, is "one of the most urgent needs of our time."[10] (Once again, this statement might well have been lost were it not, ironically, for J. Edgar Hoover, who recorded it in Einstein's dossier as more evidence of the scientist's dangerous character.)

America's postwar GNP may have passed $200 billion, but in

*When Mao Tse-tung's Red Army took power in 1949, stunned U.S. politicians traded charges and countercharges over who was responsible for "losing China." (The politicians did not explain when the United States had come to acquire China in the first place.)

the rice paddies and rubber plantations of Asia, brown-skinned people, often barefoot, carrying guns, were unimpressed. The national liberation movements, almost all battling Western colonizers, leaned more toward socialism than capitalism. While the United States often sided with the former colonial powers trying to suppress these rebellions, the Soviets (whatever Moscow's mixed motives may have been) wasted little time in providing the guerilla armies with political and often material support.

The Soviet Union was also making "an astounding comeback" from the destruction of World War II, "rebuilding its industry, regaining military strength." It was clearly emerging as the primary challenger to America's first-place position in world power-politics.[11] How to deal with Moscow became Washington's primary postwar preoccupation.

Most American policymakers, Democrats and Republicans alike, found it easier to blame the growing national liberation movements on the Russians than on British, French, or Dutch—or U.S.—colonialism. "Revolutionary movements in Europe and Asia were described to the American public as examples of Soviet expansionism—thus recalling the indignation against Hitler's aggression." Since World War II had been such a financial bonanza for American capitalism, Charles E. Wilson, president of General Motors, known for speaking bluntly,* proposed an ongoing partnership between business and the military for "a permanent war economy," with Moscow as the chief enemy.[12]

"An Iron Curtain has descended across Europe," Prime Minister Churchill, with President Truman at his side, told an audience in Fulton, Missouri, on March 5, 1946, just six months after the war's end. A month earlier, Stalin had given his "Two Camps" speech in which he argued that Western designs against the Soviet Union made peaceful coexistence virtually impossible. In Churchill's "Iron Curtain" speech, which signaled the

*A few years later, as secretary of defense under President Eisenhower, Wilson would become an embarrassing public-relations problem for the administration when he declared, "What's good for General Motors is good for America."

West's new cold war course, the British prime minister "went on to claim that no less than God himself had divined that America be the one nation to carry the responsibility of the [atomic] bomb."[13]

But not everyone jumped on the cold-war wagon. Questioning the abrupt relabeling of the Soviets from wartime ally (despite their heavy losses, the Red Army had defeated Hitler's troops at Stalingrad and then battered the German armies back to Berlin) to Red Menace, and worried about the sudden shift from a hard-won peace to a new war-footing, some, like Einstein, warned that the cold-war road would lead to annihilation. And when word leaked out that the State Department, the FBI, and other federal agencies were bringing former Nazis into the United States to help in the "war on communism," Einstein was among the protesters.*

From 1946 through 1948, American opponents of the cold war gathered around Henry Wallace, vice president (during most of World War II) under Roosevelt, and then secretary of commerce under Truman—until Truman dismissed him for being "soft on the Russians." After he was fired in September 1946, Wallace received thousands of letters and telegrams from Americans, well known and unknown, who were worried about the trend toward militarism and angry about federal inaction against lynching. Some wrote because they simply admired a man who stood by his principles. One of the earliest messages came from Einstein: "Your courageous intervention deserves the gratitude of all of us who observe the present attitude of our government with grave concern."[14]

A year later, Einstein invited Robeson to his home in Princeton. No cramped, backstage quarters this time, and no discussion (except perhaps reminiscing) of stage performances. This

*In December 1946, according to his FBI dossier, Einstein joined a group of "more than 40 scientists, educators, clergymen and other persons" who signed a protest against the granting of permanent residence and citizenship to Nazi scientists who were working for the U.S. Army; memo from Deputy Director Mickey Ladd to Hoover, January 6, 1947 (p. 632 of Correlation Summary in the FBI's Einstein dossier).

time, in October 1947, their agenda was politics. Einstein had also invited Wallace, who was about to announce his candidacy for president on the Progressive party ticket—a campaign both Einstein and Robeson supported—and another Wallace supporter, radio broadcaster Frank Kingdon. (With so many of Hoover's targets in one place, did the G-men bump into each other as they stood on the sidewalk or sat in their cars trying to be inconspicuous outside Einstein's house on Mercer Street?) The UPI photograph of that group is the only published photograph of Einstein and Robeson together. According to Einstein's FBI file, "The *Chicago Star,* dated 10/4/47, page 2, contained a photograph of EINSTEIN together with Henry A. Wallace, Dr. Frank Kingdon of the Progressive Citizens of America, and Paul Robeson. . . . An accompanying article stated that EINSTEIN had invited Wallace to his NJ home and expressed his 'great admiration for Wallace's courage and devotion to the fight for world peace.'"[15]

Besides support from Einstein and Robeson, the Wallace campaign received backing from an army of celebrities, including Helen Keller, Thomas Mann, Congressman Claude Pepper, columnist Max Lerner, and Anita McCormick Blaine, heiress to the International Harvester fortune, who sent the campaign a check for $10,000. At a conference cosponsored by the CIO and NAACP, the National Farmers Union and other liberal organizations, the delegates cabled Wallace that he had "the support of millions and millions who believe in the program of Franklin Roosevelt."[16] Most, if not all of these groups agreed that Moscow was America's main rival in the world, but they favored talking over shooting, negotiations over war.

"Coexistence or no existence" was one of Wallace's main campaign slogans. Despite accusations from the right, Wallace was not a communist—although the Communist party backed his campaign. He was neither a world government advocate like Einstein, nor pro-Soviet like Robeson, but he believed it was critical to work out peaceful cooperation with the Soviets instead of a cold war, a position Einstein and Robeson shared.

But it was in the area of civil rights that the Progressive party made its biggest impact. Besides challenging Truman to end lynching and establish equal employment practices in federal government agencies, Progressive party candidates refused to speak in any auditorium or meeting hall with segregated seating. This meant that in most southern cities they met only in black communities.

At the 1948 Democratic convention, the conservative southern Democrats split off from the party to form the Dixiecrat party, supporting Strom Thurmond's run for president. By September, Truman had adopted a liberal persona for his campaign. He also adopted or co-opted a number of planks from the Progressive party platform—most notably, ending segregation in the armed forces shortly before Election Day.

More than a million Americans voted for Wallace in 1948—the first substantial third-party showing from the left in twenty-four years (and the last in the twentieth century). Yet the number was far lower than Progressive party leaders had hoped for.* Red-baiting had taken its toll, as well as Truman's dexterity at co-opting Wallace's more liberal platform. Wallace's showing in 1948 was the high point; within a few years, both the candidate and the party became early casualties of the "Red scare," the anticommunist crusade that became known as McCarthyism.†

This was the other side of the cold war coin: if you don't support this war, you're an enemy of America. If you're not against the Russians, you're a Red—or at least a "pinko" or a "comsymp." Truman later complained about the "great wave of hysteria" sweeping the nation, but Miller and Nowak, among others, point out that he "was in large measure responsible for creating that very hysteria." On March 22, 1947, Truman issued Executive

*Wallace garnered 1,157,063 votes (2.38% of the total), less than Strom Thurmond's Dixiecrat vote of 1,169,032. His best showing was among predominantly black and Jewish voters in New York and California.

†In 1952, the Progressive party ran West Coast attorney Vincent Hallinan and African American newspaper publisher Charlotta Bass for president and vice president. They received only a fraction of Wallace's vote, and when what remained of the CP switched its strategy to working secretly within the Democratic party, the Progressive party dissolved.

Order 9835 requiring every federal employee to undergo political screening, and assigning the Department of Justice to make a list of "subversive organizations." Suddenly, millions of employees had to be screened for loyalty to the United States. With some qualms, Truman agreed to give total screening authority to J. Edgar Hoover's FBI.[17]

It would be three more years before Wisconsin's senator Joseph McCarthy and his anticommunist antics appeared on the national scene, but McCarthy's *ism* began without him.

Under Truman's loyalty program, Hoover's FBI investigated some 6.6 million people in less than six years—and came up with zero cases of espionage.[18] However, when its achievement is measured not by the number of spies uncovered but by the aura of fear created, then the Truman-Hoover loyalty program was successful indeed. More than a million employees a year were investigated about what they believe, what groups they belong to, who their friends are, and what they read. Add to that the fear created by the jailing and firing of communists and their defenders; the Taft-Hartley Act, making it a crime for union officers to be communists; loyalty oaths required for many if not most jobs, from carpenter to college teacher; and congressional committees threatening witnesses with jail if they refused to "cooperate" by naming names of friends and associates who might be leftists.

There is no doubt that part of the goal of the anticommunist crusade was to drive communists and their sympathizers out of the positions they held in unions,[19] schools, and a variety of other jobs—a goal the crusaders achieved. But the suspicion is unavoidable that part of their goal was the creation of fear itself.

If a frightened population is an easy-to-control population, it may explain why both sides in the cold war discouraged dissent and crushed questioning. At the end of World War II, the United States emerged stronger than ever, and Franklin Roosevelt's statement, "We have nothing to fear but fear itself," had never seemed truer. Yet within five years, American schoolchildren were ducking under wooden desks for protection against possible atomic bombs; otherwise intelligent citizens were spending thousands of dollars to excavate their own backyards and install concrete-lined ditches called bomb shelters; and citizens were calling the FBI to report that their neighbors harbored "communist sympathies." Those who refused to cower were attacked and, above all, isolated.

If you are Albert Einstein and feel free to criticize U.S. militarism around the world and thought-control at home, this means that Hoover's FBI (with help from several other federal agencies) will launch a top-secret, five-year campaign to discredit you, attempt to link you to Russian spies, and initiate a plan to deport you as "an undesirable alien."[20] What might be called Hoover's Get-Einstein project was launched on February 13, 1950, the morning after Einstein appeared on Eleanor Roosevelt's first nationwide television program and warned against the "concentration of tremendous financial power in the hands of the military" in America, as well as the "close supervision of the loyalty of the citizens [and] intimidation of people of independent political thinking."*

If you are Paul Robeson, this means even sharper attacks against you, as we shall see.

But to those running America's Red scare, far more important than merely attacking the resisters—which often produced only more defiance—was isolating them. With this in mind, Hoover, McCarthy, HUAC, and their cohorts made it clear to organizations like the ACLU and NAACP that it would be easy for the groups and their officers to avoid attacks—subpoenas, arrests, and Red-baiting "exposés" in the media. All they had to do was the patriotic thing: defend their nation and dissociate themselves from the Reds.

To isolate radicals, intimidate liberals. That was the game

*Einstein's warning could not have been closer to the mark. One month later, on March 19, Paul Robeson became the first American to be banned from television when NBC barred his appearance on the very same program, "Today with Mrs. Roosevelt." NBC executive vice president Charles Denny declared: "No good purpose would be served by having him speak on the issue of Negro politics"; Stewart, *Paul Robeson,* xxxiii.

plan, and with the ACLU and NAACP at least, the Hoover team won in a walk. During a seven-year period, ACLU officials and staff secretly "fed the FBI . . . internal reports, memoranda, files, minutes and other material and documents." They gave the Bureau reports "on the politics and private lives of individuals," as well as information on ACLU activities and internal discussions. Morris Ernst, ACLU general counsel and a committed anticommunist, maintained a "special relationship" with Hoover for many years, sending him a series of private memos that began: "My dear Edgar, for your eyes alone." Ernst boasted of their friendship in a *Reader's Digest* article with a surprisingly revealing title: "Why I No Longer Fear the FBI." Not surprisingly, through most of the McCarthy period, the ACLU refused to represent communists.[21]

There is no evidence that anything like the ACLU's bedding down with Hoover's Bureau occurred in the NAACP, but the historian Gerald Horne reports that during the 1950s Red scare, NAACP officials "did not hesitate to touch base with the FBI."[22]

Several historians have noted the link between McCarthyism and racism. Even as the Red scare was warming up in 1948, the NAACP complained of the "increasing tendency on the part of government agencies to associate activity on interracial matters with disloyalty."[23] The case of dismissed Labor Department employee Dorothy Baily provides a telling example. In her hearing before the Loyalty Review Board one board member asked Bailey, an African American, "Did you ever write a letter to the Red Cross [to protest] the segregation of blood?"[24]

Unfortunately, the NAACP's response to Red-baiting was, for the most part, to adapt. In 1948, a majority of the NAACP Board voted to expel W.E.B. Du Bois from the organization, despite widespread protest from members and chapters around the country. The board took action primarily because of Du Bois's support of the Wallace campaign, but clearly, too, because he refused to separate himself from Robeson and others on the left. The leadership "cracked down on the independence of local branches," and fear of communism and/or of coming under government attack was so pervasive in the NAACP that the organization kept away from even slightly Red-colored causes. President Walter White "knew that the organization was inherently vulnerable to red-baiting."[25]

Although it wasn't part of the leadership's plan, "the purges weakened the organization." The NAACP "became more moderate, more middle class, and smaller. The Detroit membership [dropped] from 25,000 during the war to 5,162 in 1952."[26]

As early as 1946, the NAACP refused to join the American Crusade to End Lynching because Robeson was co-chairman, and its board criticized Du Bois for supporting it. During the 1950s, they remained silent when the State Department refused passports to Robeson and Du Bois. Furthermore, in the many cases where African Americans faced death sentences after trials that were dubious at best—cases the black press often labeled "legal lynching"—the NAACP would not get involved, no matter how unjust the conviction, if the communist-led Civil Rights Congress (CRC) was in any way connected.

Led by William L. Patterson, an African American communist, the militant CRC combined legal advocacy and mass actions. It is perhaps best known for its detailed account of lynching in America, *We Charge Genocide,* that Patterson, Robeson, and other CRC leaders presented to the United Nations on December 18, 1951. The document charged the U.S. government with genocide against the African American population, citing thousands of lynchings and shootings by police as well as arrests, trials, and executions of black Americans on trumped-up charges. (The United States had refused to sign the U.N.'s antigenocide charter, and U.S. influence at the United Nations ensured that the CRC petition was never considered.) Besides the NAACP's aversion to Red-tainted causes, the two groups approached these cases "from two different points of view," explained Robert Harris in *The Nation*:

> The NAACP felt that the state and federal laws and legal procedures were basically just. Therefore, it relied on legal

> appeals on the hope that somewhere along the line the miscarriage of justice would be rectified. The CRC felt that the fundamental basis of the courts and the laws . . . was to preserve an unjust system of Negro oppression—a system with which the federal courts would not interfere. Thus, its approach was political, to bring the maximum public pressure to bear.[27]

As with the organization of interracial labor unions, the Communist party openly played a leadership role in such mass-protest defense campaigns, going back to the Scottsboro case in the 1930s. The clash with the NAACP notwithstanding, "The party's fight for the concrete economic needs of the unemployed and working poor, its role in organizing sharecroppers in Alabama . . . and its vigorous courtroom battles in behalf of African Americans . . . attracted a considerable section of America's black working class and intelligentsia."[28]

A number of these courtroom battles, mostly but not only in southern states, attracted media coverage during the years following World War II, especially in the black and left-wing press. Each case is different, a separate drama, usually a tragedy, with its own cast of heroes and villains. But in one respect, all the cases are the same: those charged with rape or murder are African Americans; their accusers are white.

The CRC's most famous case was the defense of a thirty-six-year-old black truck driver in Laurel, Mississippi, named Willie McGee, arrested in 1945 and charged with raping a white woman. "They say my husband raped Willametta Hawkins," Rosalie McGee told the Harlem newspaper *Freedom*. "But I say if anybody was raped, which there wasn't, it was Willametta Hawkins raped Willie, not the other way around." In a scenario recalling the false rape accusation that is at the center of *To Kill a Mockingbird*, "Willametta Hawkins wouldn't leave Willie alone. She followed him around on the job where he worked. . . . A white woman in the south if she wants a Negro man for herself she don't care if he has a wife and children or not. If he does what she wants she can holler rape; if he doesn't, she can still holler rape. If somebody else don't like what's going on, they can holler rape. No way can the Negro man win."[29]

To begin to comprehend what Willie McGee, his family, friends, and CRC attorneys went through, it is necessary to imagine *To Kill a Mockingbird* without the fair-minded, kindly judge, the balanced courtroom atmosphere, and the black spectators free from intimidation and violence. Instead, picture a courtroom with judge, jury, and spectators expecting a hanging, with armed men charging forward, guns raised, at the slightest suggestion that Ms. Hawkins might have been anything less than the perfect picture of pure, southern white womanhood.

Willie McGee's first trial lasted less than a day, and the all-white jury took two and a half minutes to reach a verdict. "It was obvious," as Jessica Mitford writes, "that a ritual race murder was in the offing."[30]

McGee lived through three trials. The reversals of his convictions were based on transparently contradictory and perjured testimony and the total exclusion of blacks from the jury pool. He had numerous stays of execution, thanks to a national and international protest campaign organized by the CRC and the tireless efforts by the group's attorneys, several of whom were physically attacked and beaten by Mississippi mobs. After one stay of execution ordered by Supreme Court justice Harold H. Burton in July 1950, "small groups of men were reported to have gathered about the courthouse building in Laurel during the afternoon, expressing resentment at Justice Burton's action and denouncing 'the damn New York Communists' and 'outside interference.'"[31]

The *Jackson Daily News*, Mississippi's largest newspaper, reflected—and at the same time, actually encouraged—the race-hatred in the white crowds milling around courthouses and squares in many of the state's cities and towns. In May 1951, after Paul Robeson publicly urged McGee's freedom, the newspaper issued a virtually unveiled call for a lynching: "Paul Robeson, Negro singer and notorious Communist, who declares he prefers Russia to the United States, blew off his loud bazoo to the effect that the next step should be to get Willie McGee out of jail. It could happen—but not in the way Robeson is thinking about."[32]

Despite Judge Burton's stay, the full Supreme Court refused—for the third time—to review Willie McGee's case. The night before he was electrocuted by the state of Mississippi, he wrote to his wife Rosalie, "Tell the people the real reason they are going to take my life is to keep the Negro down. . . . They can't do this if you and the children keep on fighting. Never forget to tell them why they killed their daddy. I know you won't fail me. Tell the people to keep on fighting. Your truly husband, Willie McGee."[33]

Einstein had protested McGee's conviction—"Any unprejudiced human being must find it difficult to believe that this man really committed [rape]"—and also attacked the death sentence. Besides Einstein, others signing on to "Save Willie McGee" included Clifford Odets, Diego Rivera, David Siquieros, Jean Cocteau, Jean-Paul Sartre, Albert Camus, Richard Wright, Dmitri Shostakovich, and Serge Prokofiev, in addition to several black-led and interracial unions in the United States, and eleven French newspapers of diverse political views.

After McGee's execution, *Time* and *Life* magazines directed angry editorial criticism—at the communists. Not a word about the case itself. But on May 14, 1951, *Time* added another editorial target:

> To Communists all over the world, "the case of Willie McGee" had become surefire propaganda, good for whipping up racial tension at home and giving U.S. justice a black eye abroad. Stirred up by the Communist leadership, Communist-liners and manifesto-signers in England, France, China and Russia demanded that Willie be freed. . . .
>
> [But] not only Communists took up the cry. In New York, Albert Einstein signed a newspaper ad protesting a miscarriage of justice.

If he had been white, it's a safe bet McGee never would have been executed. No white man has ever been sentenced to death for rape in the Deep South.

Another CRC campaign involved issues of women's rights as well as racial justice. It began on November 4, 1947:

Rosa Lee Ingram let out a scream as John Stratford knocked her down with the butt of his rifle. It was a short scream, but long enough for two of her sons, working nearby, to hear. Ingram was a forty-year-old black Georgia tenant farmer and widowed mother of twelve sons. Stratford, a neighboring white tenant farmer, had reportedly been sexually harassing her. But as Stratford moved toward her this time, her sons Wallace and Sammie arrived; one of them wrestled the gun away and struck Stratford on the head. Stratford would not harass anyone else again.

Charged with murder, Rosa Lee Ingram and her sons were convicted in a one-day trial, with a court-appointed defense attorney, by an all-white jury in Ellaville, Georgia. They were sentenced to die in the electric chair. In a hearing on the case in April 1948, their death sentences were commuted to life in prison, but the judge refused to grant a new trial or to consider the question of an all-white jury.

The CRC helped to set up three new organizations led by black women, which for the next six years waged a campaign of protest activities to free the Ingrams. These CRC-linked, radical women's organizations were "founded primarily by black women who had some association with the Communist Party," according to historian Robin D. G. Kelley. He adds that perhaps the most important of these groups was the Sojourners for Truth and Justice founded by Charlotta Bass, Shirley Graham Du Bois, Louise Thompson Patterson, Alice Childress, Beah Richards, and Rosalie McGee (Willie McGee's widow).[34] In a dramatic move, the CRC filed a petition, written by W.E.B. Du Bois, with the United Nations, arguing that the Ingram convictions violated the Universal Declaration of Human Rights approved by the U.N. General Assembly. The United Nations refused to consider the case, but the action attracted international media coverage.

"My organization is with you one hundred per cent," Ida Henderson, co-chair of the Women's Convention Auxiliary of the National Baptist Convention, told the Ingram campaign. Besides approaching churches, the protesters organized four public rallies for the Ingrams in Atlanta, collected 100,000 signatures on petitions, and sent five multiracial groups of women to Georgia over a period of five years to go door-to-door in communities and talk to people about the case. The women in the Ingram campaign also met with journalists, spoke on the radio, lobbied Georgia politicians, and enlisted support from black sororities and a number of the left-led interracial labor unions mentioned earlier. They collected money to send to Rosa Lee Ingram and her sons in jail. "Finally, the Ingrams were freed because of such tireless and selfless efforts."[35]

Not all the CRC's (or the NAACP's) defense efforts were successful, of course. As in the Willie McGee case, worldwide protests in the Martinsville Seven "rape" case could not stop the state of Virginia from proceeding with their electrocutions.*

But half a century before there was DNA testing,† the Civil Rights Congress was organizing demonstrations to protest unjust and racist court cases; sending delegations to meet with politicians, journalists and community activists; and telling the story of each case around the world, even as its lawyers challenged the convictions and sentences. The CRC also set up committees of prominent people to speak out publicly on each case. Several of these committees included Einstein. In addition, he publicly endorsed the CRC itself. (See Part II, Einstein's FBI File on Civil Rights.)

*In Mississippi and Virginia, all fifty people executed for rape in the first half of the twentieth century were black; Horne, *Communist Front?* 75.

†Since 1992, attorneys for the Innocence Project, a nonprofit group based in New York, have used "DNA fingerprinting" to reverse the convictions of 153 men (as of November 2004), including many on death row who were awaiting execution for crimes—usually rape or murder—they did not commit. The attorneys argue that even where there is no DNA evidence available, the pattern of their successful cases shows a system slanted against defendants of color and those without money.

Einstein would quite likely have endorsed an NAACP-sponsored protest committee as well, but the NAACP, as mentioned earlier, generally did not organize protests when they took on a case. While the association did direct lobbying and media outreach, it focused its efforts on trials and appeals. Nevertheless, Einstein expressed admiration for Walter White. The NAACP secretary, who was blond, blue-eyed, and very fair-skinned, might well have passed for white, but chose instead to organize and work for civil rights. In October 1947, his article, "Why I Remain a Negro," appeared in *The Saturday Review of Literature.* It prompted Einstein to send a comment:

> On reading the White article one is struck with the deep meaning of the saying: There is only one road to true human greatness—the road through suffering. If the suffering springs from the blindness and dullness of a tradition-bound society, it usually degrades the weak to a state of blind hate, but exalts the strong to a moral superiority and magnanimity which would otherwise be almost beyond the reach of man.
>
> I believe that every sensitive reader will, as I did myself, put down Walter White's article with a feeling of true thankfulness. He has allowed us to accompany him on the painful road to human greatness by giving us a simple biographical story which is irresistible in its convincing power.[36]

White's story may have been irresistible, but when it came to political causes, Einstein joined Robeson, Du Bois, and the CRC. Indeed, almost every civil rights group Einstein endorsed after 1946, including the Council on African Affairs cited earlier, had Robeson in the leadership.

Perhaps because he had seen the Nazis use the "communist" scare tactic, he did not shrink from Robeson's Red glare. Like Robeson, the CRC had close ties to the Communist party. While defending Rosa Lee Ingram, Willie McGee, the Martinsville Seven, and other African Americans they saw as victims of racist frame-ups, the CRC also supported the more than one hundred CP officials jailed under the Smith Act* during the McCarthy/Hoover period.[37] CRC statements pointed to Hitler's Germany where the Nazis had first rounded up the communists while most liberals shrugged from what they thought was a safe distance.† It was an example Einstein's memory agreed with. "The fear of communism," he declared at the height of the McCarthy era, "has led to practices which have become incomprehensible to the rest of civilized mankind."[38]

His outspokenness on civil rights included a virtually unknown 1948 interview with the *Cheyney Record*, the student newspaper of a then-small black college, Cheyney State, in Pennsylvania:

"Race prejudice has unfortunately become an American tradition which is uncritically handed down from one generation to the next," Einstein declared. That he agreed to the interview is hardly surprising, given Einstein's previous visit to Lincoln University and his openness in talking and writing to young people.[39] He continued, "The only remedies [to racism] are enlightenment and education. This is a slow and painstaking process in which all right-thinking people should take part." (For the full text of the interview, see Part II, Document H.)

Shortly after the Cheyney interview, Einstein extended his organizational network by sending a message to the "Southwide Conference on Discrimination in Higher Education" held at Atlanta University in 1950 and sponsored by the Southern Conference Educational Fund. With the Red scare, congressional investigating committees like HUAC had Red-baited virtually any southern

*The Smith Act, originally enacted in 1940, outlawed "conspiracy to teach and advocate the overthrow of the government by force and violence." During World War II, the federal government indicted several members of the Trotskyite Socialist Workers party under the Smith Act. The Communist party made no protest against those indictments.

†Not all liberals shrugged. Einstein had signed an unsuccessful appeal, along with Kathe Kollwitz and Heinrich Mann, in 1932, urging the socialists and communists to unite behind a single slate of candidates that might have been able to defeat the fascists.

group that called for integration, and driven many of them out of existence. The Highlander Folk School, where Rosa Parks took part in interracial discussions during the summer before her famous arrest for refusing to move to the back of a Montgomery, Alabama, bus, was one of the few organizations that managed to survive. Another was the Southern Conference Educational Fund (SCEF). (See Part II, Document I, for a full description.)[40]

Four years before *Brown v. Board of Education,* SCEF sponsored a rare integrated conference in the South (albeit at a black university) to oppose racism in southern universities.* In his greeting to the group, Einstein wrote:

> If an individual commits an injustice he is harassed by his conscience. But nobody is apt to feel responsible for misdeeds of a community, in particular, if they are supported by old traditions. Such is the case with discrimination. Every right-minded person will be grateful to you for having united to fight this evil that so grievously injures the dignity and the repute of our country. Only by spreading education among all of our people can we approach the ideals of democracy.
>
> Your fight is not easy, but in the end, you will succeed.[41]

Perhaps Einstein's most effective civil rights action was testimony he didn't actually give. At the start of 1951, the federal government indicted W.E.B. Du Bois, then chairman of the Peace Information Center, and four of the group's other officers for failing to register as "foreign agents." The government's principal charge was that the Peace Information Center—described by

*If you were an African American football or basketball star at a northern or western university in 1950, you would almost always be left at home when your school traveled to play a southern school. Yet signs of change were emerging. In 1955, thousands of students in Atlanta held an unprecedented demonstration, burning effigies of Georgia's governor Griffin to demand that he permit the Georgia Tech football team to play an interracial Pittsburgh team in that year's Sugar Bowl; *New York Times,* December 4, 1955, 1.

Kelley as an "antinuclear, anti–Cold War" group—had committed the "overt act" of circulating the Stockholm Peace Petition.*

If one needs a single image to represent McCarthyism in America, it might well be the picture of W.E.B. Du Bois in 1951 facing a judge in a federal courtroom—the world-renowned black scholar, at the age of eighty-three, goateed, short in height but standing unbent, wearing a pinstriped, three-piece suit and handcuffs.[42] Like Robeson, Du Bois had refused to go along with Washington's anti-Soviet, anticommunist policies, refused to cooperate with congressional investigating committees, had his passport suspended, and had been ousted from the NAACP.

Shortly after the federal indictment, Einstein sent Du Bois a copy of his just-published book, *Out of My Later Years.* It was almost exactly twenty years after Einstein had first heard from Du Bois and written his statement for *The Crisis.* In April, Du Bois wrote back and included information about his upcoming court case: "Mrs. Du Bois and I have received your autographed book with deep appreciation and will read it with pleasure and profit. I am venturing to enclose with this letter a statement on a case in which you may be interested."

Einstein quickly volunteered to testify as a defense witness in Du Bois's federal trial. To give Einstein's appearance in court the maximum impact, defense attorney Vito Marcantonio† held

*Several million people signed the worldwide peace petition initiated in 1950 by the Stockholm-based, pro-Soviet World Peace Council. The petition declared: "We demand the outlawing of atomic weapons as instruments of intimidation and mass murder of people. We demand strict international control to enforce this measure. We believe that any government which first uses atomic weapons against any other country whatsoever will be committing a crime against humanity and should be dealt with as a war criminal. We call upon all men and women of goodwill throughout the world to sign this appeal." HUAC denounced it as "the most extensive piece of psychological warfare ever conducted on a world scale . . . a smoke screen for [Communist] aggression"; Fariello, *Red Scare,* 486.

†Marcantonio was a popular, independent, fiery left-wing congressman from New York's East Harlem, supported by both the Italian and Puerto

back the announcement until the last minute. In a rare, firsthand account, Shirley Graham Du Bois describes the judge's response:

> The prosecution rested its case during the morning of November 20. . . . Marcantonio . . . told the judge that only one defense witness was to be presented, Dr. Du Bois. [But] Marcantonio added casually to the judge, "Dr. Albert Einstein has offered to appear as a character witness for Dr. Du Bois." Judge [Matthew F.] McGuire fixed Marcantonio with a long look, and then adjourned the court for lunch. When court resumed, Judge McGuire . . . granted the motion for acquittal.[43]

Confronted with the prospect of international publicity that would have resulted from Einstein's testimony, the judge dismissed the case for lack of evidence before the defense had a chance to present its witnesses.

Nine days later, Du Bois wrote to Einstein again:

> My dear Dr. Einstein:
> I write to express my deep appreciation of your generous offer to do anything that you could in the case brought against me by the Department of Justice.
>
> I am delighted that in the end it was not necessary to call upon you and interfere with your great work and needed leisure, but my thanks for your generous attitude is not less on that account.
>
> Mrs. Du Bois joins me in deep appreciation.
>
> Very sincerely yours,
> W.E.B. Du Bois[44]

(See full correspondence, Part II, Document K.)

Rican communities there. Among his many distinctions, he was the only member of Congress to vote against sending U.S troops into the 1950 Korean "police action."

CHAPTER 10

Einstein and Robeson, II

Einstein and Robeson met for a last time in October 1952 when Einstein again invited Robeson to his Princeton home. Their meeting went almost completely unreported. Even J. Edgar Hoover didn't know about it, as Einstein had sent the invitation to Robeson through a mutual friend, assuming—correctly—that the FBI was intercepting mail and phone calls.[1] It was the height of the Red scare.

Fear of a foreign, evil empire was the rationale for the Red scare, but to be successful every witch hunt needs local witches, and Robeson was one of the highest on Hoover's witch-list. Defiantly leftist, Robeson continued both to assail racial injustice in America and express friendship for the Soviet Union. In September 1949, a mob attacked his picnic-concert at Peekskill, New York, stoning and beating picnickers, both black and white, screaming for Robeson's blood as state police watched and, in some cases, joined the mob's attack. The attack was not a massacre only because thousands of labor union members and World War II veterans who had volunteered to defend the picnic formed a human wall around the grounds. The self-defense group had no weapons other than an occasional baseball bat (for the game they hoped to hold after the concert).

While the Peekskill attack has been reported, it is less well known that, under cover of the screaming mob, an attempt was made to assassinate Robeson. The self-defense group also sent scouts out into the surrounding woods before the concert began. At one point, "They flushed up two local patriots who had made a little nest for themselves up there overlooking the valley. And they had high-powered rifles with telescopic sights. In other words, they want to kill Paul."* The first two snipers were chased off, but to protect against others, an interracial group of tall, broad-shouldered veterans volunteered to stand behind Robeson as he sang "with the full knowledge that they were providing a barrier of human flesh between him and [possible] snipers."[2]

Cars throughout Westchester County displayed bumper stickers declaring: "Wake Up America, Peekskill Did," and "Communism Is Treason, Behind Communism Stands—the Jew!" Earlier that year, Einstein had written that America was "half-fascistic."[3]

After Peekskill, the U.S. State Department—probably on orders from Hoover[4]—canceled Robeson's passport, blocking him from accepting concert invitations from around the world. In just a few blacklisted years, he was barred from almost every concert hall in the country—even black churches received threatening phone calls when they invited Robeson, and some canceled his performances in fear they would lose their insurance. Robeson's annual income plummeted from $100,000 to $6,000. Despite or because of that atmosphere, Einstein invited Robeson to visit him again.

The two men spent the entire afternoon together, probably close to six hours. "The conversation continued until it started to get dark outside," according to Lloyd Brown, Robeson's close friend and colleague, who accompanied him to Einstein's house. When they arrived, Helen Dukas ushered them upstairs to Einstein's working room that had been renovated so that half of its

*Under the headline, "Was FBI Involved in Attempts to Kill Paul Robeson?" an article by J. J. Johnson in the newspaper the *Daily World*, October 25, 1979, reported that Paul Robeson's son, Paul Jr., after reviewing some 3,500 documents in the FBI's file on his father, found that "there were at least four occasions, in 1946, 1955, and twice in 1958, when cars in which his father was being transported were sabotaged. Both the cars and drivers in each case were under FBI surveillance"; Davis, *Paul Robeson Research Guide*, 636, and *Daily World*, 11.

back wall was a huge picture window overlooking a garden.* "Einstein was reclining on a sofa with a fly-swatter in his hand," Brown reported, "And when we walked in he waved the fly-swatter and said, 'This is my war.'" One of Brown's stories about that session, quoted often, is worth repeating here: At one point when Robeson left the room, Brown, nervously trying to make conversation, said, "Dr. Einstein, it's an honor to be in the presence of a great man." Einstein frowned, clearly annoyed, and replied, "But you came in with a great man."[5]

Understanding what happened at that meeting between Einstein and Robeson is key to appreciating the bond between the two men. Brown's account gives us some idea of the content of their discussion: they talked about everything from music (Einstein regretfully said he was no longer able to play the violin) to why Einstein had recently declined an invitation to become president of Israel: "Einstein said he had been opposed to the Jewish State in the beginning. He had been in favor of a bi-national state. He said, when you have a state then you get an army and he was sorry the bi-national state didn't happen. He said he had written about this and it was in his book of essays, *Out of My Later Years*."[6] In response to Einstein's questions, Robeson reported on his efforts to resist McCarthyism at home, as well as the intensifying freedom struggle in South Africa. Einstein said he thought the South Africans should use Gandhi's tactics of passive resistance, and—eight years before the student sit-ins launched the civil rights movement—Einstein argued that black Americans, too, should adopt nonviolence as their strategy.

Fascinating as the conversation must have been, some meetings mean more than their agendas, and this was one of them. The only two public accounts of that session—a column by Robeson in the Harlem-based newspaper *Freedom* and Brown's

*The room had floor-to-ceiling bookcases on two sides with a desk in the middle, overlooking the garden. Einstein once said of the room, "I do not actually feel as being within a building"; Bucky, *Private Albert Einstein*, 42.

interview on Gil Noble's television program, *Like It Is*[7]—emphasize the importance to Robeson of Einstein's support at a time when the singer was under the sharpest political attack. "Dr. Einstein . . . expressed warm sympathy for my fight for the right to travel," Robeson reported in his column two weeks after their meeting. In Brown's words, "Einstein's invitation was a definite act of solidarity, especially coming after Peekskill."[8]

But something else at least equally important may have occurred that October afternoon in Princeton. If Einstein had simply wanted to make a showing of solidarity, as the host, especially given his poor health, he could have ended the meeting gracefully after an hour. That he didn't do so is a clue to how much the session with Robeson meant to him.

At the Institute for Advanced Study, the great scientist's thirty-year search for a unified field theory (linking the forces of gravity and electromagnetism to one underlying force) continued, so far without the positive results he hoped for. But he was used to pitting his stubborn curiosity against the unknown. It was the state of the world, not the universe, that aggravated Einstein. As the planet's richest, most powerful country, sweat-drenched in Red-scare hysteria, rushed to build bigger, better, and deadlier bombs, Einstein's politics became increasingly anti. From the house on Mercer Street, his interviews and articles warned of potential catastrophe and even "annihilation."[9]

August of 1952 found Einstein in a state of serious gloom—deepening for nearly three years—over the political situation in America and the parallels he saw to Germany just a few decades earlier. "You are very right in assuming that I am badly in need of encouragement," he wrote in November 1950 to a friend in Indiana. "Our nation has gone mad. America's obsession with anti-Communism and the rising tide of political fear reminds me of the events in Germany." Two months later, he wrote: "The dear Americans have vigorously assumed [the Germans'] place. . . . The German calamity of years ago repeats itself. . . . People acquiesce and align themselves with the forces of evil, and one stands by, powerless." And in another letter to friends in Europe: "I hardly ever felt as alienated from people as right now. . . . The worst is that nowhere is there anything with which one can identify. Everywhere brutality and lies." And again on January 31, 1951: "Honest people constitute a hopeless minority."[10]

One significant event on May 8, 1951, must have deepened his depression: Willie McGee was electrocuted by the state of Mississippi. As we have seen, McGee had been on death row for nearly six years while the worldwide campaign to save him had won reversal of two convictions and several stays of execution.

The following month, Einstein wrote: "Those who propagandize against the alleged external enemy have won the support of the masses." By the start of 1952, he seemed to have almost lost hope: "Saddest of all is the disappointment one feels over the conduct of mankind in general." And in September, just a few weeks before Robeson's visit, he wrote to a friend in Italy about his sense of loneliness and isolation.[11]

It is a safe bet that Einstein's support meant a great deal to the besieged Robeson. But the afternoon they spent together in October 1952 may have meant even more to Einstein. Did Einstein confide his sense of hopelessness to Robeson? He probably didn't need to. "Dr. Einstein asked me about my life today as an artist," Robeson wrote in his account of their conversation. We can imagine Einstein's eyes widen as Robeson described his most recent concert. Five months earlier, after the government had barred him from going to Canada, where he had been invited by that country's mineworkers union, he sang at the Peace Arch spanning the border between Washington State and British Columbia, Canada. Some 40,000 union members turned out on the Canadian side and cheered, as Robeson, standing on the U.S. side, sang into a microphone that sent his magnificent voice—and his defiance—across the barred border.[12]

We don't know what spurred Einstein's return from despair in 1953, but Robeson's visit might well have helped. Six hours discussing strategies for social change with a colleague Einstein admired, receiving reports about struggles around the world, and hearing the great artist, activist, and friend describe the response to each new, vindictive government attack—more protest rallies, concerts, and organizing—could not have hurt Einstein's morale. In any case, there is no doubt that by the beginning of 1953, he no longer felt powerless. In January, he spoke out against the death sentence for the Rosenbergs, and a few months later publicly challenged Senator McCarthy, calling on witnesses to refuse to cooperate with congressional investigators, even if it meant going to jail. His challenge to the inquisitors (Einstein's word) made the front page of the *New York Times* twice in six months. On June 12, reporting on his letter to Brooklyn schoolteacher William Frauenglass, the *Times* headline read:

'REFUSE TO TESTIFY,' EINSTEIN ADVISES INTELLECTUALS CALLED IN BY CONGRESS

Since the *Times* was a media agenda-setter, then as now, Einstein's defiance became national and international news.[13]

Senator McCarthy, predictably, called him "an enemy of America," but Einstein's friend Erich Kahler observed that after Einstein's decision "to stir the conscience of the public," he underwent a dramatic mood change: "I have never seen him so cheerful and sure of his cause."[14]

Einstein and Robeson had their differences, to be sure,* including, significantly, their estimates of the Soviet Union. Robeson loyally supported Moscow while Einstein felt the Stalin regime

*While they both viewed the congressional investigations as "witch hunts," and "inquisitions," they differed on how to resist. Einstein believed taking the Fifth—refusing to answer a question to avoid self-incrimination—made a witness seem guilty. He thought witnesses should simply refuse to answer improper questions, without using "the well-known subterfuge of . . . the Fifth Amendment." Robeson followed the Communist party line that taking the Fifth was the surest way of staying out of jail, since those who relied on other defenses (such as the First Amendment) had been prosecuted for contempt of Congress and jailed.

had become an anti-Semitic dictatorship. But both clearly believed that what they shared was more important. And, as one uncovers more about their friendship, it is surprising to learn *how much* they had in common. These two giants—who belong to the world, but also to Witherspoon Street—were not only friends, but shared a passionate belief in the brotherhood and sisterhood of human beings, in music as a great international unifier, and in the right and responsibility of all of us to resist fascists and lynchers everywhere. They both believed also that it is possible to have a more equitable society than one where private profit is God.

But perhaps most important—and it may well be they came to understand this during that afternoon they spent together in 1952—they shared the belief that when it comes to social justice, you don't have to win, but you do have to keep struggling. It was a sentiment Einstein expressed a year before he died when he said the struggle for human rights "is an eternal struggle in which a final victory can never be won. But to tire in that struggle would mean the ruin of society." It was also a commitment Robeson immortalized when, singing at a mass rally for the antifascist forces in Spain, he changed the line from "Old Man River" from "I'm tired of livin' but scared of dyin'" to "I must keep fighting until I'm dying."[15]

The party also argued (as did Supreme Court Justice William O. Douglas) that the Fifth Amendment was an important constitutional guarantee against tyranny. But Einstein remained unconvinced. "Taking the Fifth [appears to be an] admission of guilt," he told the *Newark News* (December 17, 1954).

CHAPTER 11

"My Friend, Doctor Einstein"

Two days after Einstein's death on April 18, 1955, a headline in the *Daily Princetonian* declared:

NEIGHBORING KIDS UNFOLD LEGEND
ABOUT THEIR FRIEND MR. EINSTEIN

Einstein's long and friendly relationship with the children of Princeton was the subject of this piece—from a young girl who claimed to give him gumdrops for help with her arithmetic, to three boys who in the summer would meet Dr. Einstein for water gun battles and bring an extra water gun for him. However one special child, an African American child, was described as "the Doctor's most constant friend." In 1955 he was eleven years old; his name was Harry Morton and, as readers were told, he was from Battle Road:

> I'd be doing homework and he'd [Einstein] come by everyday around four all alone, with his dachshund just shuffling along with his feet dragging, looking down at the ground. So I would run out and meet him and we walked and talked and traded jokes. I hardly ever understood them. . . . They were all about arithmetic and things, but I laughed anyway.

Harry would often visit the scientist at home, the *Princetonian* report continued. Einstein "showed him his chemistry equipment in the cellar and let him stay until supper, while he worked at his desk. 'He always wrote for about ten minutes with funny numbers, then he got up and walked around the desk twice.

Once in a while he got real mad.'" According to Evelyn Turner, a neighborhood friend who looked after Harry Morton:

> Harry lived with us for about four years when he was in the second, third, fourth, and possibly fifth grades. While he didn't talk with me directly about Einstein, I remember Harry talking with the other boys about how he had met him.
>
> I remember that Harry was a very bright boy, very talkative, very inquisitive, and he had no fear of strangers. He would visit his mother at the Aydelottes' home on weekends and go to their church on Sunday mornings, too. Every Sunday night, his mother would drop him off at our house. She thought it was better for Harry to be raised in a place where there were other children his own age.

Harry Morton passed away in 1992. Joi Morton, his wife, had these recollections:

> To Harry, Einstein was just a man that he knew. My husband never played this up. My daughter asked me how her father and Einstein got to be friends. Harry's mother worked as a domestic for Dr. Frank Aydelotte who was then head of the Institute for Advanced Study. Their house sat on the property of the Institute—88 Battle Road. You used to be able to come down Battle Road and drive onto the Institute. Einstein used to go by the house every day, well every few days—and he and Harry would go for walks. As I've told my daughter, knowing your father, he would be right there when Einstein walked his dog. This was almost every day for three or four years. Later, when the Mortons moved to Clay Street, Einstein still came by. I think Harry was eleven when Einstein died.
>
> Harry told me he used to walk with Einstein. I could not imagine it. Him walking with Albert Einstein—and he's telling me they were very good friends. I was like, yeah all right—it was not registering with me. It wasn't until his

mother verified what he was saying that it became real to me. His mother worked for the Aydelottes. They stayed in the same house when the next family, the Lilienthals, moved in.

Not only did Harry have a brilliant vocabulary, but from a young age, he had that way of being able to talk to anybody and had a deep understanding of things way beyond his years.

Harry would talk about some racist things in Princeton. Jackson Street was the dividing line. It later became Paul Robeson Place. That's where the black community started. From there on down to Woodbridge Avenue that's where 90 percent of the black people lived. Blacks didn't do a whole lot of venturing out. We're talking about a lot of brilliant people who didn't even try to go to Princeton University. There was an invisible line that they didn't cross. They wanted Harry to apply to Princeton, but he didn't want to do it.

Harry would go over to Einstein's house sometimes. He would try to get answers from him for his homework, but Einstein couldn't get Harry's homework. He would go down in Einstein's basement to see the chemistry equipment. Harry accepted him as a really nice old man.

Alice Satterfield commented, "You didn't have to be a scientist to be invited into Einstein's house. He was just very down to earth. Too bad they made that absent-minded image of him."

PART II

DOCUMENTS

CHAPTER 1

Einstein's Statements on Race and Racism

Einstein's antiracist statements and articles are brought together here for the first time. Some have never been published before, others rarely.

DOCUMENT A.

"To American Negroes," *The Crisis,* February 1932, with letters from and an editor's note by W.E.B. Du Bois

October 14, 1931
Mr. Albert Einstein
Haberlandstrasse 5
Berlin, W. 30, Germany

Sir:
I am taking the liberty of sending you herewith some copies of THE CRISIS magazine. THE CRISIS is published by American Negroes and in defense of the citizenship rights of 12 million people descended from the former slaves of this country. We have just reached our 21st birthday. I am writing to ask if in the midst of your busy life you could find time to write us a word about the evil of race prejudice in the world. A short statement from you of 500 to 1,000 words on this subject would help us greatly in our continuing fight for freedom.

With regard to myself, you will find something about me in "Who's Who in America." I was formerly a student of Wagner and Schmoller in the University of Berlin.

I should greatly appreciate word from you.

Very sincerely yours,
W.E.B. Du Bois

Einstein replied on October 29, 1931:

Mr. W. E. Burghardt du Bois
Editor, "The Crisis"
69 Fifth Avenue
New York, NY

My Dear Sir!
Please find enclosed a short contribution for your newspaper. Because of my excessive workload I could not send a longer explanation.

With Distinguished respect,
Albert Einstein

TO AMERICAN NEGROES

A Note from the Editor [Dr. Du Bois]:
The author, Albert Einstein, is a Jew of German nationality. He was born in Wurttemburg in 1879 and educated in Switzerland. He has been Professor of Physics at Zurich and Prague and is at present director of the Kaiser-Wilhelm Physical Institute at Berlin. He is a member of the Royal Prussian Academy of Science and of the British Royal Society. He received the Nobel Prize in 1921 and the Copley Medal in 1925.

Einstein is a genius in higher physics and ranks with Copernicus, Newton and Kepler. His famous theory of Relativity, advanced first in 1905, is revolutionizing our explanation of physical phenomenon and our conception of Motion, Time and Space.

But Professor Einstein is not a mere mathematical mind. He is a living being, sympathetic with all human advance. He is a brilliant advocate of disarmament and world Peace and he hates race prejudice because as a Jew he knows what it is. At our re-

quest, he has sent this word to THE CRISIS *with "Ausgezeichneter Hochachtung" ("Distinguished respect"):*

It seems to be a universal fact that minorities, especially when their individuals are recognizable because of physical differences, are treated by majorities among whom they live as an inferior class. The tragic part of such a fate, however, lies not only in the automatically realized disadvantages suffered by these minorities in economic and social relations, but also in the fact that those who meet such treatment themselves for the most part acquiesce in the prejudiced estimate because of the suggestive influence of the majority, and come to regard people like themselves as inferior. This second and more important aspect of the evil can be met through closer union and conscious educational enlightenment among the minority, and so emancipation of the soul of the minority can be attained.

The determined effort of the American Negroes in this direction deserves every recognition and assistance.

—Albert Einstein

In response, Dr. Du Bois wrote again:

November 12, 1931
Professor Albert Einstein
Berlin, W. 30
Germany

My dear Professor Einstein:
I am under deep obligations to you for the statement which you sent us. I know how busy you must be and it is exceedingly kind for you to take time for writing this. I shall take the liberty of sending you a copy of my magazine each month.

Very sincerely yours,
W.E.B. Du Bois

"To American Negroes," *The Crisis* 39 (February 1932), 45. Thanks to the Crisis Publishing Co., Inc., the publisher of the magazine of the National Association for the Advancement of Colored People, for the use of this material. This article was also published as "On Minorities" in *Mein Weltbild,* in 1934, and reprinted in *Ideas and Opinions,* in English, in 1950. Reissued in 1982 by Crown Publishers, Inc. Used by permission of Crown Publishers, a division of Random House, Inc. Correspondence between Du Bois and Einstein: MS 312, Special Collections and Archives, W.E.B. Du Bois Library, University of Massachusetts at Amherst, reproduced courtesy of the library and David Du Bois. Einstein's note to Du Bois was translated from the German by Rebecka Jerome.

DOCUMENT B.
Address at the Inauguration of the "Wall of Fame" at the World's Fair in New York, 1940

It is a fine and high-minded idea, also in the best sense a proud one, to erect at the World's Fair a wall of fame to immigrants and Negroes of distinction.

The significance of the gesture is this: it says: These, too, belong to us, and we are glad and grateful to acknowledge the debt that the community owes them. And [focusing on] these particular contributors, Negroes and immigrants, shows that the community feels a special need to show regard and affection for those who are often regarded as step-children of the nation—for why else this combination?

If, then, I am to speak on the occasion, it can only be to say something in behalf of these step-children. As for the immigrants, they are the only ones to whom it can be accounted a merit to be Americans. For they have had to take trouble for their citizenship, whereas it has cost the majority nothing at all to be born in the land of civic freedom.

As for the Negroes, the country has still a heavy debt to discharge for all the troubles and disabilities it has laid on the Negro's shoulders, for all that his fellow-citizens have done and to some extent still are doing to him. To the Negro and his wonderful songs and choirs, we are indebted for the finest contribution in the realm of art which America has so far given to the world. And this great gift we owe, not to those whose names are engraved on this "Wall of Fame," but to the children of the people, blossoming namelessly as the lilies of the field.

In a way, the same is true of the immigrants. They have contributed in their way to the flowering of the community, and their individual striving and suffering have remained unknown.

One more thing I would say with regard to immigration generally: There exists on the subject a fatal miscomprehension. Unemployment is *not* decreased by restricting immigration. For [unemployment] depends on faulty distribution of work among those capable of work. Immigration increases consumption as much as it does demand on labor. Immigration strengthens not only the internal economy of a sparsely populated country, but also its defensive power.

The Wall of Fame arose out of a high-minded ideal; it is calculated to stimulate just and magnanimous thoughts and feelings. May it work to that effect.

Einstein Archives, Box 36, file 28-529, 1–2. Reproduced courtesy of the Albert Einstein Archives, The Jewish National & University Library, The Hebrew University of Jerusalem, Israel, and Princeton University Press.

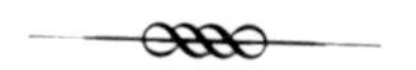

DOCUMENT C.
"The Negro Question," *Pageant*, January 1946

I am writing as one who has lived among you in America only a little more than ten years, and I am writing seriously and warningly. Many readers may ask: "What right has he to speak out about things which concern us alone, and which no newcomer should touch?"

I do not think such a standpoint is justified. One who has grown up in an environment takes much for granted. On the other hand, one who has come to this country as a mature person may have a keen eye for everything peculiar and characteristic. I believe he should speak out freely on what he sees and feels, for by so doing he may perhaps prove himself useful.

What soon makes the new arrival devoted to this country is the democratic trait among the people. I am not thinking here so much of the democratic political constitution of this country, however highly it must be praised. I am thinking of the relationship between individual people and the attitude they maintain toward one another.

In the United States everyone feels assured of his worth as an individual. No one humbles himself before another person or class. Even the great difference in wealth, the superior power of a few, cannot undermine this healthy self-confidence and natural respect for the dignity of one's fellow-man.

There is, however, a somber point in the social outlook of Americans. Their sense of equality and human dignity is mainly limited to men of white skins. Even among these there are prejudices of which I as a Jew am clearly conscious; but they are not important in comparison with the attitude of the "Whites" toward their fellow-citizens of darker complexion, particularly toward Negroes. The more I feel an American, the more this situation pains me. I can escape the feeling of complicity in it only by speaking out.

Many a sincere person will answer me: "Our attitude towards Negroes is the result of unfavorable experiences which we have had by living side by side with Negroes in this country. They are not our equals in intelligence, sense of responsibility, reliability."

I am firmly convinced that whoever believes this suffers from a fatal misconception. Your ancestors dragged these black people from their homes by force; and in the white man's quest for wealth and an easy life they have been ruthlessly suppressed and exploited, degraded into slavery. The modern prejudice against Negroes is the result of the desire to maintain this unworthy condition.

The ancient Greeks also had slaves. They were not Negroes but white men who had been taken captive in war. There could be no talk of racial differences. And yet Aristotle, one of the great Greek philosophers, declared slaves inferior beings who were justly subdued and deprived of their liberty. It is clear that he was enmeshed in a traditional prejudice from which, despite his extraordinary intellect, he could not free himself.

A large part of our attitude toward things is conditioned by opinions and emotions which we unconsciously absorb as children from our environment. In other words, it is tradition—besides inherited aptitudes and qualities—which makes us what we are. We but rarely reflect how relatively small compared with the powerful influence of tradition is the influence of our conscious thought upon our conduct and convictions.

It would be foolish to despise tradition. But with our growing self-consciousness and increasing intelligence we must begin to control tradition and assume a critical attitude toward it, if human relations are ever to change for the better. We must try to recognize what in our accepted tradition is damaging to our fate and dignity—and shape our lives accordingly.

I believe that whoever tries to think things through honestly will soon recognize how unworthy and even fatal is the traditional bias against Negroes.

What, however, can the man of good will do to combat this deeply rooted prejudice? He must have the courage to set an example by word and deed and must watch lest his children become influenced by this racial bias.

I do not believe there is a way in which this deeply entrenched evil can be quickly healed. But until this goal is reached there is no greater satisfaction for a just and well-meaning person than the knowledge that he has devoted his best energies to the service of a good cause.

Pageant, January 1946. Reprinted in *Out of My Later Years* (New York: The Philosophical Library, 1950), a collection of Einstein's essays that he selected himself. A Citadel Press Book. Reprinted by arrangement with Kensington Publishing Corp. *www.kensingtonbooks.com.*

DOCUMENT D.
Speech to Lincoln University Students and Faculty, May 3, 1946

The only possibility of preventing war is to prevent the possibility of war. International peace can be achieved only if every individual uses all of his power to exert pressure on the United States to see that it takes the leading part in world government.

The United Nations has no power to prevent war, but it can try to avoid another war. The U.N. will be effective only if no one neglects his duty in his private environment. If he does [neglect it], he is responsible for the death of our children in a future war.

My trip to this institution was in behalf of a worthwhile cause. There is separation of colored people from white people in the United States. That separation is not a disease of colored people. It is a disease of white people. I do not intend to be quiet about it.

The situation of mankind today is like that of a little child who has a sharp knife and plays with it. There is no effective defense against the atomic bomb. . . . It can not only destroy a city but it can destroy the very earth on which that city stood.

Excerpts from the *Baltimore Afro-American,* May 11, 1946. Reprinted by permission of the Afro-American Newspapers Archives and Research Center.

DOCUMENT E.
Letter to President Harry S. Truman on Antilynching Law, September 1946

September 1946
To: the President of the United States
Honorable Harry S. Truman
The White House
Washington, D.C.

May I wholeheartedly endorse the aims of this delegation, in the conviction that the overwhelming majority of the American people is demanding that every citizen be guaranteed protection against acts of violence. To insure such protection is one of the most urgent tasks for our generation. A way always exists to overcome legal obstacles whenever there is a determined will in the service of such a just cause.

Yours respectfully,
Albert Einstein

Supporting the protest by the American Crusade to End Lynching that Einstein cochaired with Paul Robeson. Most of the letter was quoted in the *New York Times*, September 23, 1946. Reproduced courtesy of the Albert Einstein Archives, The Jewish National & University Library, The Hebrew University of Jerusalem, Israel, and Princeton University Press.

DOCUMENT F.
Message to the National Urban League Convention, September 16, 1946

Dr. Lester B. Granger
Executive Director
National Urban League
1133 Broadway
New York City

My dear Dr. Granger:
I greet your assembly in the conviction that it deals with one of the most important problems of this country. The contrast among groups is a constant threat which imperils minorities. This threat becomes more acute in times of economic stress and insecurity. However, the defeat of this threat is of importance not only for the minorities but for the country as a whole. For an unleashed war of the groups against each other is the sure way to the loss of civil rights. "Divide and Rule" has always been the maxim of tyrants and betrayers of the people.

Pessimists have often claimed that mutual hostility among groups is unavoidable, because violence, distrust and lust for power are indestructible and powerful characteristics of human nature that unceasingly influence the actions of men. No man of sound feeling and judgment is deluded by such an argument. Men are also greedy—yet we have achieved a state in which stealing is relatively rare. Every disease of society can be overcome if there is the firm will for a cure in the people.

The worst disease under which the society of our nation suffers, is, in my opinion, the treatment of the Negro. Everyone who is not used from childhood to this injustice suffers from the mere observation. Everyone who freshly learns of this state of affairs at a maturer age, feels not only the injustice, but the scorn of the principle of the Fathers who founded the United States that "all men are created equal." He feels that this state of affairs is unsound in a country which in many other things is justly proud of a high degree of development. He cannot understand how men can feel superior to fellow-men who differ in only one point from the rest: They descend from ancestors who, as a protection against the destructive action of the radiation of the tropical sun, gained a more strongly pigmented skin than those whose ancestors lived in countries farther from the equator.

One can hardly believe that a reasonable man can cling so tenaciously to such prejudice, and there is sure to come a time in which school-children in their history lessons will laugh about the fact that something like this did once exist.

Today, however, this prejudice is still alive and powerful. The fight against it is difficult as is the fight on every issue in which the enemy has thoughtlessness and a fatal tradition on its side. What is to be done?

There is a sound feeling of justice in the people; if we succeed in putting this feeling to the service of our cause, then the goal will be achieved. First we must make every effort [to insure] that the past injustice, violence and economic discrimination will be made known to the people; the taboo, the "let's-not-talk-about-it" must be broken. It must be pointed out time and again that the exclusion of a large part of the colored population from active civil rights by the common practices is a slap in the face of the Constitution of the nation.

We must strive [to ensure] that minorities be protected against economic and political discrimination as well as against attack by libelous writings and against the poisoning of youth in the schools. These endeavors are important, but not as important as the intellectual and moral enlightenment of the people. For example, it is clear that in the South, through economic pressure on the colored population, wages in general are kept down, and that thereby the majority of the people [white and black] suffers pauperization through reduced buying power. If the majority knew of the root of this evil, then the road to its cure would not be long.

In our times, the consideration of the influence all things exert on world politics plays a special role. The Fathers of our Constitution achieved a world-wide political effect by stating and realizing just principles that found enthusiastic agreement among good men everywhere. In our times, in which circumstances have placed great influence on international affairs into the hands of the United States, this nation could for a second time be the source of health and liberation, if we learned to base our influence not on battleships and atomic bombs but on being a shining example in our internal affairs and on liberating creative ideas on social and world affairs. But without a just solution of the racial—and more generally the minority—problem, our example cannot be considered shining.

One more remark. We have become accustomed to judge everything according to tangible and measurable viewpoints. Food in calories, national and private income in dollars, ownership or non-ownership of a bathtub and of a water-closet. In accordance with this attitude I have spoken so far only of such tangible things as economic and political discrimination. These are really important. Yet there are other things which are just as decisive for the happiness or unhappiness of the human being. A man may own everything that puts him on a high level according to a crudely materialistic point of view, and have a miserable existence. He may be persecuted by fear, hatred or envy, he may be deaf to merry songs and blind to blooming life. It is similar for social groups, even for whole nations.

There is a thing which one could call the moral climate of a society. In one society, there may be a preponderance of distrust, malice and ruthless egotism; in another, the enjoyment of beauty and of the blooming life, compassion for the suffering of one's fellow-men and rejoicing in their happiness. This moral climate of the society to which we belong, is of decisive influence on the value of life for each of us and it cannot be understood through the tables of statistics of economists or in any scientific way.

One thing is certain: No mechanism can give us a good moral climate as long as we have not freed ourselves from the prejudices to whose defeat you are devoting yourselves.

Yours very sincerely,
Albert Einstein.

Einstein Archives, file 57–543. Reproduced courtesy of the Albert Einstein Archives, The Jewish National & University Library, The Hebrew University of Jerusalem, Israel, and Princeton University Press.

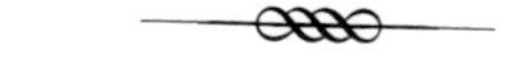

DOCUMENT G.
On Walter White, October 1947

Walter White, secretary of the National Association for the Advancement of Colored People, was blond, blue-eyed, and very fair-skinned and might well have passed for white, but chose instead to organize and work for civil rights. In October 1947, his article, "Why I Remain a Negro," appeared in *The Saturday Review of Literature*. Einstein sent the following letter to the editors, which was published on November 11, 1947:

> On reading the White article one is struck with the deep meaning of the saying: There is only one road to true human greatness—the road through suffering. If the suffering springs from the blindness and dullness of a tradition-bound society, it usually degrades the weak to a state of blind hate, but exalts the strong to a moral superiority and magnanimity which would otherwise be almost beyond the reach of man.
>
> I believe that every sensitive reader will, as I did myself, put down Walter White's article with a feeling of true thank-

fulness. He has allowed us to accompany him on the painful road to human greatness by giving us a simple biographical story which is irresistible in its convincing power.

Albert Einstein

The Saturday Review of Literature, November 11, 1947. Reprinted in *Albert Einstein, The Human Side,* ed. Helen Dukas and Banesh Hoffman (Princeton, N.J.: Princeton University Press, 1979). Reprinted by permission of Princeton University Press.

DOCUMENT H.
Interview with the *Cheyney Record,* October 1948

In October 1948, Einstein agreed to an interview with the *Cheyney Record,* the student newspaper of Cheyney State Teachers College, a black college in Pennsylvania (now Cheyney University).

Q: Do you feel that the scientists who gave us the atomic bomb should be held morally responsible for any destruction wrought by the bomb?

A: No. It is true that advances in physics have made possible the application of scientific discoveries for technical and military purposes, which engenders great danger. The responsibility, however, lies with those who make use of these new discoveries rather than with those who contribute to the progress of science—with the politicians rather than the scientists!

Q: Do you feel that race prejudice in the United States is merely a symptom of a world-wide conflict?

A: Race prejudice has unfortunately become an American tradition which is uncritically handed down from one generation to the next. The only remedies are enlightenment and education. This is a slow and painstaking process in which all right-thinking people should take part.

Q: Can mathematics be a tool for the solving of social problems as well as scientific considerations?

A: Mathematics is a useful tool for social science. In the actual solution of social problems, however, goals and intentions are the dominant factors.

Q: Do you feel that democracy can always solve the problems of society?

A: Democracy, taken in its narrower, purely political, sense suffers from the fact that those in economic and political power possess the means for molding public opinion to serve their own class interests. The democratic form of government in itself does not automatically solve problems; it offers, however, a useful framework for their solution. Everything depends ultimately on the political and moral qualities of the citizenry.

Q: Do you feel that a United States of Europe will solve the problem of war?

A: Creation of a United States of Europe is an economic and political necessity. Whether it would contribute to a stabilization of international peace is hard to predict. I believe yes rather than no.

The interview appeared in the *Record*'s February 1949 edition. Einstein's close friend, Otto Nathan (also the executor of Einstein's estate), reported that only three of the questions and answers were actually published. No copy of the February 1949 *Record* has been located. These questions and answers appear in Otto Nathan and and Heinz Norden, eds., *Einstein on Peace* (New York: Simon and Schuster, 1960), 501–502. Reproduced courtesy of the Albert Einstein Archives, The Jewish National & University Library, The Hebrew University of Jerusalem, Israel.

DOCUMENT I.
Message to the Southwide Conference on Discrimination in Higher Education, Sponsored by the Southern Conference Educational Fund (SCEF), Atlanta University, 1950

Established in New Orleans in 1946, SCEF's stated goal was to eradicate segregation in the South. Its membership and leadership included white and black southerners. Aubrey Williams, former head of Roosevelt's National Youth Administration, was the first president of SCEF and remained so until the mid-1960s. Eleanor Roosevelt actively supported SCEF and helped buffer it against Red-baiting attacks from racist senators like Bilbo of Mississippi and Talmadge of Georgia. After the *Brown* decision in 1954, SCEF worked closely with Martin Luther King Jr.'s Southern Christian Leadership Conference (SCLC). In the 1960s, under the leadership of Anne and Carl Braden, SCEF allied itself with the Student Nonviolent Coordinating Committee (SNCC).

> If an individual commits an injustice he is harassed by his conscience. But nobody is apt to feel responsible for misdeeds of a community, in particular, if they are supported by old traditions. Such is the case with discrimination. Every right-minded person will be grateful to you for having united to fight this evil that so grievously injures the dignity and the repute of our country. Only by spreading education among all of our people can we approach the ideals of democracy.
>
> Your fight is not easy, but in the end, you will succeed.

Einstein's 1950 greetings were quoted in the *New York Times* and published by SCEF the following year in *Discrimination in Higher Education*

(New Orleans: SCEF, 1951). The history of SCEF comes from Irwin Klibaner's article in Mary Jo Buhl, Paul Buhl, and Dan Georgakas, eds., *Encyclopedia of the American Left* (Urbana: University of Illinois Press, 1992), 736–737.

DOCUMENT J.
Interview with Peter A. Bucky

Einstein: The longer I live in America, the more sad I feel about this situation. I have spoken with many people who have told me that they had bad feelings against Negroes because of unfavorable experiences that they had by living side by side with them. I've also been told that Negroes are not equal in intelligence or in their sense of reliability and that they are irresponsible.

But I think that there is a certain amount of selfishness in this belief. By that I mean that American ancestors took these black people forcibly from their homes so that the white man could more easily acquire wealth. By suppressing and exploiting and degrading black people into slavery the white man was able to have an easier life. I really think that it is as a result of a desire to maintain this condition that modern prejudices stem.

Bucky: Is there any relation between this anti-Negro sentiment and anti-Semitism?

Einstein: Only that it is part of the continuing story of man's inhumanity. Look, as far back as Greek times, people kept slaves. The only difference then was that the slaves were white people and therefore could not be [attacked] due to racial differences. And yet, the Greek philosophers declared these slaves to be inferior. They too, were deprived of their liberty. However, being a Jew myself, perhaps I can understand and empathize with how black people feel as victims of discrimination.

Bucky: What do you think can be done in the long run to solve the problem?

Einstein: Well, there is no magic solution. I would only hope that where there is a will there is a way. I think probably that Americans will have to realize how stupid this attitude is and how harmful it is, also, to the standing of the United States. After all, every country is supposed to be looking up to this country. But I think that if individuals are really honest with themselves about this problem, they would undoubtedly recognize how wrong this bias really is.

From Peter A. Bucky, *The Private Albert Einstein* (Kansas City, MO: Andrews and McMeel, 1992), 46–47.

DOCUMENT K.
Letters from W.E.B. Du Bois, 1951

The renewed Einstein–Du Bois contact began early in 1951, when Einstein sent Du Bois a copy of his just-published book, *Out of My Later Years.* In April, Du Bois wrote back and included information about his federal indictment.

April 20, 1951

Dr. Albert Einstein
112 Mercer Street
Princeton, New Jersey

My dear Dr. Einstein:
Mrs. Du Bois and I have received your autographed book with deep appreciation and will read it with pleasure and profit.

I am venturing to enclose with this letter a statement on a case in which you may be interested. We are sir,

Very sincerely yours,
W.E.B. Du Bois

After Einstein volunteered to be a defense witness and the case against Du Bois was subsequently dismissed by federal judge Mathew F. McGuire, Du Bois wrote again.

November 29, 1951

Mr. Albert Einstein
112 Mercer St.
Princeton, New Jersey

My dear Dr. Einstein:
I write to express my deep appreciation of your generous offer to do anything that you could in the case brought against me by the Department of Justice.

I am delighted that in the end it was not necessary to call upon you and interfere with your great work and needed leisure, but my thanks for your generous attitude is not less on that account.

Mrs. Du Bois joins me in deep appreciation.

Very sincerely yours,
W.E.B. Du Bois

Du Bois correspondence: MS 312, Reel 66, no. 591, Special Collections and Archives, W.E.B. Du Bois Library, University of Massachusetts at Amherst, reproduced courtesy of the library and David Du Bois.

CHAPTER 2

From Einstein's FBI File: On Civil Rights

AMERICAN CRUSADE TO END LYNCHING

The ACEL items in Einstein's FBI file begin with a report from Army Intelligence (G-2), described by the FBI as "a completely reliable source."[1]

> When in Washington, the delegation planned to call on the White House and national figures to demand action by the administration. A parade was scheduled to be led by colored and white veterans who were to march to the Lincoln Memorial where a national religious ceremony would be held and persons who escaped lynch mobs were to be presented. . . . Dr. Albert Einstein was scheduled to appear.

As with most of Einstein's political activities, the FBI's reports on ACEL rely heavily on news stories and other published material:

> *The Philadelphia Inquirer* . . . dated 9/23/46 . . . stated EINSTEIN wrote a letter to President HARRY S. TRUMAN assailing lynching. This letter was to be delivered to President TRUMAN by a group headed by PAUL ROBESON. The *People's Voice* dated 10/5/46 . . . stated in part that EINSTEIN and PAUL ROBESON were co-chairmen of the ACEL.

The three-point program for the ACEL included: (1) apprehension and punishment of all lynchers; (2) passage of a

> Federal anti-lynching law; (3) removal of [Mississippi] Senator Theodore Bilbo. Dr. Albert Einstein has been selected as co-chairman with Paul Robeson of the Crusade.
>
> A report in the September 14, 1946 issue of the *Baltimore Afro-American* states that a one hundred-day "Crusade to End Lynching" would begin in Washington, DC with a huge rally on September 23 . . . in the afternoon, delegations will be appointed to confer with government officials [about] definite programs to stop lynching.
>
> [The ACEL pamphlet states] "Lynching cannot be permitted to continue unchecked in a nation founded upon . . . justice and equality. . . . America must rid itself of the crime of unpunished lynching. . . ." Signers, "morally indignant over this lawless violence," include ALBERT EINSTEIN.

Although Hoover frequently asserted that the Bureau was nothing more or less than a fact-collection agency,* the FBI's report on the American Crusade to End Lynching declared:

> In view of some of the endorsers, this Crusade has all the earmarks of another Communist attempt to instill racial agitation.
>
> The above information is evaluated in this report as having been received from a usually reliable source whose information was probably true.†

*In 1953, Hoover told the *New York Times*: "It is the ironclad practice of the FBI never to evaluate any of the information it receives from its own investigators or any other source"; *New York Times*, March 29, 1953.

†"No matter how much Hoover claimed his was only a fact-finding agency [of] strictly investigative nature . . . the FBI during the McCarthy era became an infiltrative organization, gathering 'intelligence'. . . more likely to compromise, embarrass and ruin those investigated than merely supply 'facts'"; Nash, *Citizen Hoover,* 106.

A subsequent entry on the ACEL states:

> Paul Robeson, who has a long record of Communist affiliations, was the moving spirit in what was known as the American Crusade to End Lynching and organized a pilgrimage to Washington, D.C. for 9/23/46. This venture was actively supported by the Communist press.

In a harbinger of hundreds, if not thousands, of FBI reports to come, one entry on the ACEL concludes by labeling the group with a tag-line that summarizes the Bureau's view of civil rights protests generally: "*RE: Foreign Inspired Education among the American Negroes,* Internal Security-C [for Communist]."

DEFENSE OF THE SCOTTSBORO NINE

> On July 5, 1931 . . . EINSTEIN, THOMAS MANN, LION FEUCHTWANGER and several others [formed] a "German committee" . . . in support of "DREISER's committee" . . . to save nine Negroes at Scottsboro, Alabama, from the electric chair.

"Dreiser's committee" refers to Nobelist author Theodore Dreiser whose "subversive" activities included defense of the Scottsboro defendants. The last of the Scottsboro defendants to go free was Hayward Patterson, who escaped to Michigan in 1948 after serving seventeen years in Alabama prisons on what the world knew were false charges. In 1950, Michigan's governor G. Mennen Williams refused to extradite Patterson back to Alabama.

CAMPAIGN TO SAVE SAM BUCKHANNON

> The *Daily Worker* of October 2, 1943, carried an article entitled "Final Hearing Set on Buckhannon Fight" with the subheading "Einstein Joins Defense to Halt Negro Extradition." The article stated in part that Professor Albert Einstein, the world renowned scientist, was the latest to add his name to those battling to save Sam Buckhannon, 34-

year-old Negro who had served fourteen years on the chain gang [in Georgia] for stealing a package of cigarettes . . . and that the National Federation for Constitutional Liberties was conducting the fight to save Buckhannon.

After months of protests, a New Jersey judge freed Buckhannon (*New York Times*, October 9, 1943).

THE DEFENSE OF WILLIE MCGEE

The *Daily Worker* of 3/27/51 described WILLIE McGEE as a Mississippi Negro victim of a rape frame-up who was seeking an appeal of a death sentence before the US Supreme Court.

According to the Worker of 4/22/51 . . . EINSTEIN stated in part: "In the face of the evidence, any unprejudiced human being must find it difficult to believe that this man really committed the rape of which he has been accused. Moreover, the punishment must appear unnaturally harsh to anyone with any sense of justice." [Informant blacked out] advised that WILLIE McGEE was a Negro convicted of rape and executed in the State of Mississippi for this crime.

THE DEFENSE OF THE "TRENTON SIX"

The "Trenton Six" refers to the trial and conviction of six Negroes accused of killing William Horner, a Trenton, NJ merchant. According to [informant's name blacked out], the defendants were represented by the Civil Rights Congress and the case has been of interest to the Communist Party of New Jersey.

The *Worker* dated 6/5/50 . . . : "EINSTEIN and 14 other individuals denounced the court attempts to deprive the Trenton Six of defense by naming attorneys of their own choice and advocated that the public help in the struggle to preserve the Bill of Rights."

DEFENSE FUND FOR NEGRO CITIZENS

A letter from the National Association for the Advancement of Colored People, Legal Defense and Educational Fund . . . dated August 14, 1943, addressed to Mr. [Ernest] Hemingway . . . enclosed a statement signed by more than one hundred leaders in almost every field of American life, an appeal to the addressee to contribute to the establishment of a vitally necessary Defense Fund to safeguard the rights of the Negro citizens.

The signers of this petition included Albert Einstein.

CIVIL RIGHTS CONGRESS (CRC)

[An article in] the *Daily Worker* dated 12/15/48 . . . entitled "Crusade to Capitol on 1/18 Will Urge Civil Rights Laws," [stated] the CRC had announced . . . that thousands of people would come to Washington . . . in a gigantic "freedom crusade" to demand payment of [Truman's] election campaign promissory notes on Civil Rights Legislation. . . . Sponsors of the "Freedom Crusade" included EINSTEIN and others.

NATIONAL COMMITTEE TO OUST BILBO

[Informant's name blacked out] furnished the Bureau with literature distributed by the National Committee to Oust Bilbo, sponsored by the Civil Rights Congress. . . . Included in this material was a letter dated Dec. 4, 1946, signed by Quentin Reynolds and Vincent Sheehan. . . . The names of 55 members of this Committee were set out including that of Albert Einstein.

THE FREEDOM CRUSADE

The Civil Rights Congress announced . . . that thousands of people would come to Washington on January 18 [1949] to demand payment of the election campaign promissory notes on Civil Rights Legislation. . . . The conference would be held at the AFL Laborer's auditorium in Washington. . . . Sponsors of the conference and the "Freedom Crusade" included Dr. Albert Einstein and others.

SOUTHERN CONFERENCE ON HUMAN WELFARE

The *Southern Patriot* of March, 1950 . . . announced that the Southern Conference Educational Fund [will hold] a southwide conference on discrimination in higher education in Atlanta University, April 8. . . . The New Jersey sponsors of the conference [included] the name of Albert Einstein.

COUNCIL ON AFRICAN AFFAIRS (CAA)

According to the *Daily Worker* dated 4/22/47 in an article captioned, "EINSTEIN says Liberation of Colonies Urgently Needed," Professor EINSTEIN sent greetings to Max Yergan, Executive Director of the Council on African Affairs on 4/21/47 [and] said in part "no reliable or lasting peace will be possible without the political and economic emancipation of the subdued and exploited African and colonial people."

DINNER TO HONOR W.E.B. DU BOIS

Counter Attack, a weekly newsletter published by the American Business Consultants, Inc. of New York City, on February 16, 1951 stated . . . that accused "Foreign Agent Du Bois" would be honored at a hotel banquet; that Dr. Du Bois' "long record of pro-Communist activities had not deterred approximately 200 people (referred to as 'notables' in Communist Party press) from tendering him a banquet in honor of his 83rd birthday"; that the dinner was scheduled to be held at the Essex House in New York City on February 23.

Counter Attack stated further that the "notable" sponsors included Dr. Albert Einstein and others.

The paper "Freedom". . . dated February 1951 . . . speaks of the Du Bois testimonial dinner: "More than 200 prominent individuals from all sections of the US, among them

Dr. Albert Einstein, Mrs. Mary McLeod Bethune, Dr. Kirtley Mather and Paul Robeson [are sponsoring the dinner] to honor Dr. W.E.B. Du Bois on his 83rd birthday . . . in the Colonnades Ballroom of the New York Essex House on February 23."*

*When the Essex House cancelled the dinner, it was held at Small's Paradise, a Harlem nightclub.

Notes

Preface

1. Two welcome exceptions, neither one an Einstein biography, have been published in the past few years: the UPI photo of Einstein with Robeson, Henry Wallace, and Frank Kingdon, which appears in this book, was included in *Albert Through the Looking Glass,* edited by Ze'ev Rosenkranz, published by the Einstein Archives, Hebrew University, Jerusalem, and reprinted in this country by Johns Hopkins University Press as *The Einstein Scrapbook* (the photo caption describes Robeson as "the baritone and civil rights campaigner"); and in *Expanded Quotable Einstein* and *The New Quotable Einstein,* Alice Calaprice added references to Robeson and the American Crusade to End Lynching. Neither of these volumes mentions W.E.B. Du Bois, the Civil Rights Congress, or Witherspoon Street.

2. Fred Jerome's interview with insists-on-anonymity museum curator, April 6, 2000.

3. Herbert Aptheker, *Anti-Racism in U.S. History,* xiii.

Chapter 1. Escape from Berlin

1. Early attacks on Einstein by German nationalists, see Otto Friedrich, *Before the Deluge,* 215.

2. Thomas Levenson, *Einstein in Berlin,* 403–404.

3. Philipp Frank, *Einstein: His Life and Times,* 231.

4. Levenson, *Einstein in Berlin,* 388.

5. Einstein "was anxious": Norman Bentwich, *My 77 Years,* cited by Ronald W. Clark, *Einstein: The Life and Times,* 508.

6. "perpetrate a murder": Antonina Vallentin, *The Drama of Albert Einstein,* 203, cited by Levenson, *Einstein in Berlin,* 414.

7. Death threat: General Hans von Seeckt, commander in chief of the German Army, described in Vallentin, *Einstein, a Biography,* 202–209.

8. "My permanent home will still be in Berlin": Einstein quoted in the *New York Times,* October 16, 1932, cited by Denis Bryan, *Einstein, A Life,* 237.

9. Abraham Pais, *Einstein Lived Here,* 190.

10. Frank, *Einstein: His Life and Times,* 231.

11. E. Krieck, cited in ibid., 228.

12. The earth literally shook as Einstein made his announcement. Evelyn Seeley reported in the *New York World Telegram* of March 11, 1933, that as he walked away from the interview at Caltech, "Dr. Einstein felt the ground shaking under his feet. Los Angeles was being visited by the worst earthquake [until that date] in its history."

13. Not Yet Hanged: Jamie Sayen, *Einstein in America,* 17; ransacked summer house in Caputh: "Nazis Hunt Arms in Einstein Home," *New York Times,* March 21, 1933, 10.

14. C. P. Snow, "Reminiscences," in A. P. French, ed., *Einstein: A Centenary Volume,* 5.

15. Correspondence between Du Bois and Einstein: W.E.B. Du Bois Papers, Archives, Library of the University of Massachusetts at Amherst.

16. Einstein's note to Du Bois translated from the German by Rebecka Jerome.

17. Albert Einstein, "To American Negroes," *The Crisis* 39 (February 1932), 45.

18. "Einstein Hails Negro Race," *New York Times,* January 19, 1932, 23.

Chapter 2. "Paradise"

1. "puny demigods on stilts": letter to Belgium's Queen Elisabeth, November 20, 1933. Einstein Archives, 32-369, cited in Alice Calaprice, *The Expanded Quotable Einstein,* 56. In another reference to Princeton's "society," Einstein wrote: "Here, the people who compose what is called 'society' enjoy even less freedom than their counterparts in Europe. Yet they seem unaware of this restriction since their way of life tends to inhibit personality development from child-

hood": letter to Queen Elisabeth, November 20, 1933, cited in Nathan and Norden, eds., *Einstein on Peace,* 245; "banishment to paradise": Sayen, *Einstein in America,* 64.

2. "the chaotic voices of human strife barely penetrate": letter to Queen Elisabeth, March 20, 1936, Einstein Archives, Princeton University (hereafter Einstein Archives).

3. William K. Selden, "Princeton University's National Historic Landmark," 9–10.

4. Princetoniana committee home page http://alumni.princeton.edu/~ptoniana/belcher.asp.

5. Ibid.

6. Ibid.

7. Woodrow Wilson was the keynote speaker at Princeton's sesquicentennial (150th year) celebration, "Princeton in the Nation's Service." http://etc.princeton.edu/campusWWW/companion/princeton_in_nations_service.

8. Albert Marrin, *The War for Independence, The Story of America,* 112.

9. Ibid., 116.

10. Ibid.

11. Leila Amos Pendleton, *A Narrative of a Negro* (Washington, D.C.: Press of R. L. Pendleton, 1912), 106.

12. Alfred Hoyt Bill, *The Princeton Campaign, 1776–1777* (Princeton, N.J.: Princeton University Press, 1948), 112.

13. Princeton University, "The Presidents of Princeton," 1991 edition—Twentieth Century, part 1, http://www.princeton.edu/pr/facts/presidents/02.htm.

14. Gerald Breese, *Footprints on Edgehill Street,* 46.

Chapter 3. The Other Princeton

1. Interview with Doris Burrell, November 30, 1995, Brooklyn Historical Society, copy courtesy of Princeton Historical Society.

2. Bruce Wright, *Black Robes, White Justice,* 36–37.

3. Seeley G. Mudd Library, Princeton University, "Slavery at Princeton," www.princeton.edu/mudd/news/faq/topics/slavery.shtml.

4. The Historical Society of Princeton, "A Community Remem-

bers—African American Life in Princeton," www.princetonhistory.org/aalife.

5. Lloyd Brown, *The Young Paul Robeson,* 22.

6. Library of Congress, Civil War and Reconstruction, 1761–1877, "African American Soldiers during the Civil War—The Negro as Soldier," http://memory.loc.gov/ammen/ndlpedu/features/timeline/civilwar/aasoldrs/nsoldier.htm.

7. Ibid.

8. Connie Escher, "Betsey Stockton," *Princeton History* 10 (1991), 77.

9. Ibid., 75.

10. *The Daily Princetonian,* January 16, 1885, 211.

11. Dwandalyn Reece King, "Completing the Historical Record: Princeton's First Exhibition on the African-American Community," *Princeton History* 14 (1997), 44.

12. Ibid., 41–42.

13. John Frelinghuysen Hageman, *History of Princeton and Its Institutions* (1879), 2: 210.

14. *New York Times,* January 3, 1903; African American files, Seeley Mudd Library, Princeton University.

15. Ibid.

16. "Klanvocation" in Hamilton: *The Trentonian* on-line by Jon Blackwell, with photo; "unmasked in the shadow of the Capitol": *Life* magazine's *A Century of Change,* ed. Richard B. Stolley, 216. Photo shows KKK parade in Washington, D.C.; more than 1,100 lynchings: August Meier and Elliott Rudwick, *From Plantation to Ghetto* (New York: Hill and Wang, 1970), cited in Martin Bauml Duberman, *Paul Robeson,* 4.

17. Daniel J. Loeb, *From Sambo to Superspade,* 34.

18. Herbert Aptheker, *A Documentary History of the Negro People,* 806.

19. Paul Robeson, *Here I Stand,* 17.

20. *Princeton Herald,* September 25, 1942.

21. The black community's self-imposed curfew: Emma Epps interview, the Princeton History Project, the Princeton Historical Society, Oral History Transcripts, Box 2, Folder 7; "drunken rich": Robeson interview in *The Daily Princetonian,* 1950 ("Paul Robeson, Born in Town; Ardently Dislikes Princeton"), Princeton Historical So-

ciety's Vertical File; "Oh, how I remember those bells!" Emma Epps in the *Trenton Gazette,* April 16, 1976 ("Remembering the Trolley, the Curfew and Kid Green's Car").

22. Message to the Urban League, see Part II, Document F.

Chapter 4. Witherspoon Street

1. Princeton Group Arts: "Rex Goreleigh: Memories of the Man and His Early Rise in Civil Rights Here," *Town Topics,* November 5, 1986, 14.

2. Marian Anderson: "complete artistic mastery": *The Daily Princetonian,* April 17, 1937; "stayed at Einstein's house": Margot Einstein quoted by Sayen, *Einstein in America,* 221.

3. Marian Anderson, *Lord What a Morning,* 267.

4. Freeman Dyson's story about Griggs restaurant: E-mail to Fred Jerome, August 18, 2004.

5. "A Community Remembers—African American Life in Princeton," The Historical Society of Princeton, http://www.princetonhistory.org/aalife/.

Chapter 5. Einstein and Robeson, I

1. "a quarter-century earlier" would have brought Einstein to Princeton in 1908, by which time the Robesons had moved to nearby Somerville. They moved there in 1906, when Paul was eight years old. But it seems highly likely that young Paul would have returned to Princeton frequently during his preteen years to visit and play ball with friends and cousins in his old neighborhood, especially considering the predominantly white composition of Somerville.

2. *New York Tribune,* November 22, 1917.

3. "spiritually located in Dixie": Duberman, *Paul Robeson,* 5; "Georgia plantation town": Wright, *Black Robes, White Justice,* 33.

4. "How rich in compassion!": Robeson, *Here I Stand,* 15; "much more communal": ibid., 10–11; "closest of ties": ibid., xi.

5. Dorothy Butler Gilliam, *Paul Robeson, All-American,* 20.

6. 1930 concert tour of Europe: "Robeson, the great Negro artist, who has so sweet a compassion for the underdogs of the world . . . [was] literally mobbed at the station in Glasgow by autograph seekers;

Robeson—in Dublin; Robeson—in Marseilles; Robeson—in Moscow; Robeson—arriving in Barcelona, going on to Madrid—singing in the American Hospital at Villa Paz, at the hospital base at Bennicasime. . . . In Oslo, Copenhagen, Stockholm, he received tumultuous, unprecedented receptions which became anti-fascist demonstrations. In Oslo, after a concert during which ten thousand people were *outside* the hall, the Nordic patriots fell on their knees, kissing his hands while tears ran down their cheeks. Robeson! A myth—a legend!" *Jewish Chronicle*, London, April 8, 1930, reprinted in Lenwood G. Davis, ed., *A Paul Robeson Research Guide*, 352.

7. Duberman, *Paul Robeson*, 132.

8. Book burning: William Shirer, *The Rise and Fall of the Third Reich*, 241. Among many descriptions of the Unter den Linden scene: Ronald W. Clark, *Einstein: The Life and Times*, 571; Otto Friedrich, *Before the Deluge*, 385; and Frank, *Einstein: His Life and Times*, 237.

9. "staring out into the darkness": Marie Seton's diary and Robeson's interview with *Berliner Zeitung*, June 21, 1960, both quoted by Duberman, *Paul Robeson*, 184–185.

10. Robeson's description of meeting with Einstein: conversations with Lloyd Brown, October-November 1998. Robeson also cited the Einstein meetings in his monthly column, "Here's My Story," in the Harlem-based newspaper *Freedom*, November 1952.

11. Prewrite and review of Robeson's 1935 concert: *The Daily Princetonian*, October 31 and November 1, 1935.

12. *The Daily Princetonian*, April 20, 1955, 3.

13. Robeson's McCarter performance dates: Dan Bauer of the McCarter Theatre, confirmed by reports in *The Daily Princetonian*.

14. While Robinson didn't join the Brooklyn Dodgers until 1947, he was signed by the Dodgers organization in 1946, amid much publicity, and played his first season for the Dodgers' farm team in Montreal.

15. The McCarter was built in the prosperous 1920s with money from Thomas N. McCarter, class of 1888, as a permanent home for the Princeton University Triangle Club, a drama/musical comedy club with "a longstanding tradition of blending topical humor with collegiate irreverence and playfulness" (Triangle Club Website). The theater opened its doors on February 21, 1930, and was financially

and politically independent from the university until 1950, when the university took title to the building and bottom-line control of the theater. Don Marsden, the Triangle Club's graduate secretary, writes: "Triangle has a long tradition of tolerance toward anyone with talent, so the integration of McCarter's stage and seating isn't really a surprise. During the mid to late 1950s, while the University was bathed in bad publicity because of anti-Semitism at the Prospect Street eating clubs, Triangle had many Jewish members (e.g., Oscar-winning screenwriter Robert S. "Bo" Goldman '53, who was elected Triangle president as a senior)": E-mail letter from Marsden to Jerome, June 28, 2004.

16. Margaret Walker quoted by Duberman, *Paul Robeson,* 263.

17. Quoted in Jeffrey C. Stewart, *Paul Robeson, Artist and Citizen,* 121.

18. Princeton was still Princeton: Duberman, *Paul Robeson,* 658, n. 9.

19. Guernica bombing: Peter Jennings and Todd Brewster, *The Century,* 198. Also see Herschel B. Chipp, *Picasso's Guernica.*

20. "The Lincolns came from all walks of life": Sam Sills's essay, "The Abraham Lincoln Brigade," in Buhle et al., eds., *Encyclopedia of the American Left,* 2–4. The Lincoln Brigade's black commander, Oliver Law, a thirty-three-year-old army veteran from Chicago, was reportedly popular because he took part in every activity he assigned to his men. Law was killed in battle: Steve Nelson, *Steve Nelson, American Radical,* 205–218.

21. Attacks on Einstein over his support for the Spanish Republic and Lincoln Brigades, Rankin from white Mississippi: *Detroit Times,* October 28, 1945. HUAC reports, January 3, 1940, and March 29, 1944, also quoted in the right-wing Catholic newspaper *The Tablet,* February 25, 1950.

22. At least 60 percent of Americans in the Lincoln Brigades were communists, with most of the others socialists or members of other left groups: Sills, "Abraham Lincoln Brigade," 3. Sills also cites casualties among the Lincolns.

23. Among the thousands of scientists supporting the antifascist fight in Spain: Nobel Prize–winning chemist Harold Urey, geneticist L. C. Dunn, zoologist Selig Hecht, engineer Walter Rautenstrauch,

astronomer Harlow Shapley, physicists Arthur Compton and J. Robert Oppenheimer, pathologist John Peters, mathematician Dirk Struik; leading figures in American anthropology, Franz Boas and Ruth Benedict; and physiology, Walter Cannon and A. J. Carlson. In England, the long list of antifascist scientists was headed by the noted crystallographer and political activist J. D. Bernal.

24. African American support for Spanish Republic: Robin D. G. Kelley, in *Encarta Africana*, on-line newsletter.

25. "The artist must take sides": Quoted in the *Washington Tribune*, December 4, 1937; Davis, *A Paul Robeson Research Guide*, 220.

26. "I do not want my children to become slaves": Robeson interview with Guillén, first printed in *Mediodía*, Havana, 1938, reprinted in *Bohemia*, Havana, May 7, 1976; Philip Foner, ed., *Paul Robeson Speaks*, 122–123.

27. Robeson's impact on the men in Spain: Interview with British volunteers, George Baker and Tommy Adlam, by Anita Sterner for a 1978 BBC program on Robeson; Duberman, *Paul Robeson*, 218.

28. "I have never met such courage in a people": Robeson interview with Guillén, cited above, Foner, *Paul Robeson Speaks*, 125; "We don't feel deeply enough": Duberman, *Paul Robeson*, 220.

29. "ashamed the Democratic nations had failed": Einstein quoted in the *New York Times*, February 5, 1937, 5; cited in several sections of the FBI's Einstein file, including the Summary Report in section 8.

30. Einstein's message to New York mass meeting on April 18, 1937, in Nathan and Norden, *Einstein on Peace*, 274.

31. Casals tribute: French, *Einstein, A Centenary Volume*, 43.

Chapter 6. "Wall of Fame"

1. Einstein's view of the war as a struggle between slavery and self-determination, and Einstein to Soviet ambassador Maxim Litvinov on postwar world government: Nathan and Norden, *Einstein on Peace*, 320; Robeson believed the Russians had eradicated prejudice within a single generation: Duberman, *Paul Robeson*, 282.

2. So many Americans: A small number of Einstein's former colleagues in the War Resisters League stuck to their pacifist antiguns and went to jail rather than to war, but the vast majority of Ameri-

cans, including Einstein, believed that armed force was necessary in order to defeat Hitler; Madison Square Garden rally: *New York Times*, June 23, 1942, 1. Also see Duberman, *Paul Robeson*, 253.

3. Hoover's prewar ties to Nazi police officials; his invitation to Himmler to attend the international police conference: FBI File 65-3598; his reception of Himmler's aide and correspondence with German officials: Anthony Summers, *Official and Confidential*, 134; also see reference to Hoover's letter to KRIPO chief counsel W. Fleischer: Frank J. Donner, *The Age of Surveillance*, 86; postwar Nazi links: *U.S. Intelligence and the Nazis* (Washington, DC: Nazi War Crimes and Japanese Imperial Government Records Interagency Working Group of the National Archives and Records Administration, 2004); see especially chapter 9, "What the FBI Knew." For a discussion of Hoover's secret ties to isolationists in Congress, including Senator Burton Wheeler of Montana, see Jerome, *The Einstein File*, 46–47.

4. Low-profile mini witch hunts: In one wartime project, the FBI and HUAC harassed and monitored a group of left-leaning German writers who had come to America to escape Hitlerism. The agencies used the war with Germany as a pretext to spy on these *anti*fascist Germans. In his book (*"Communazis"*), Alexander Stephan cites hundreds of government files to expose how, during the World War II years, the FBI and other agencies targeted and harassed German refugee writers such as Bertoldt Brecht, Thomas and Heinrich Mann, Lion Füchtwanger, and Erich Remarque, not because they were suspected backers of Hitler, but because they were known antifascists, often with communist contacts or socialist ideas. Einstein was in touch with many of these artists, shared their hatred for Nazism, and with some, such as Thomas Mann, shared a role in several political protests—and FBI reports.

The antileft surveillance policy was not limited to refugee writers. Throughout the war years, the FBI continuously spied on American writers and artists the Bureau viewed as "Red" or leaning in that direction. For example, one confidential FBI memo to Hoover in 1943 warned that Ernest Hemingway had been "active in aiding the Loyalist cause in Spain [and] his views are 'liberal' and . . . he may be inclined favorably to Communist political philosophies." A year later, as American troops were landing in Normandy to fight the Nazis, a memo on "Communist influence or control" in the Writers War Board (a government agency headed by Rex Stout, which had begun as a conservative, anticommunist group, but become increasingly involved in the antifascist war effort) stated: "It is significant to note that the name of Langston Hughes appears with the advisory council. . . . Hughes, you will recall, is the Negro Communist poet famous for the Communistic, atheistic poem, 'Goodbye Christ.'" FBI memo on Hemingway from Mickey Ladd to Hoover: Herbert Mitgang, *Dangerous Dossiers,* 44–45; memo on Hughes, also from Mickey Ladd to Hoover, ibid., 179.

5. "powerful national leaders": Theoharis, *J. Edgar Hoover, Sex, and Crime,* 164. President Franklin Roosevelt's memos to and meetings with Hoover, authorizing the FBI to investigate communists (as well as Nazis) are detailed in a number of Hoover biographies listed in the Bibliography, especially those by Gentry and Theoharis and Cox. Hoover's personal connection to the power elite and his closest friend outside the FBI was George E. Allen, a financier and lobbyist. Allen served as a director of numerous manufacturing and insurance companies, including Republic Steel, and was a friend of Presidents Roosevelt, Truman, and Eisenhower. He played poker regularly with Truman, who appointed him to the Reconstruction Finance Corporation, and later became Eisenhower's bridge and golf partner. His unique brand of modesty is revealed in the title of his autobiographical book, *Presidents Who Have Known Me.*

6. Denied security clearance: Jerome, *The Einstein File,* chap. 4, "Banned from the Bomb."

7. Surveillance of Robeson: Duberman, *Paul Robeson,* 253–254, and Kenneth O'Reilly, *"Racial Matters,"* 31.

8. *Kristallnacht*: *New York Times,* November 11, 1938, 1.

9. "willing to assume responsibility": Richard Rhodes, *The Making of the Atomic Bomb,* 305; also cited by Spencer Weart and Gertrude Szilard, *Leo Szilard: His Version of the Facts,* 83.

10. "My dear Professor," Roosevelt's reply to Einstein, October 19, 1939: Nathan and Norden, *Einstein on Peace,* 297.

11. Most of Einstein's "I Am an American" interview, NBC Radio, June 22, 1940, is reprinted in ibid., 312–314.

12. "Germany had just invaded Poland": essay by Gilda Snow in Jennings and Brewster, *The Century*, 211.

13. Among the 600 names inscribed on the twenty-one panels of the Wall of Fame, the total number of American Indians listed was 4. The number of Negroes was 42. The remaining 554 honorees came from 59 countries, including 79 from Germany, 74 from England, and 1 from China: "WALL OF FAME of The American Common, World's Fair of 1940 in New York," published by the American Common, Robert D. Kohn, Chairman (also vice president of the World's Fair).

14. "Wall of Fame" speech: Einstein Archives, Box 36, file 28-529, 1–2.

15. Roosevelt's abstention "disastrous," Einstein's letter to his friend and fellow Nobelist Harold Urey, August 16, 1940: Nathan and Norden, *Einstein on Peace*, 315–317. Einstein said Roosevelt's support for England was "halfhearted and insufficient instead of unconditional solidarity."

16. Cheered America's entry into the war, a message from Einstein to the German people, December 7, 1941: Nathan and Norden, *Einstein on Peace*, 320.

17. "lies in the overthrow of fascism": Robeson press conference, quoted in the *Labor Herald*, September 25, 1942, 1, cited in Davis, *A Paul Robeson Research Guide*, 180.

18. "I don't think there will be trouble": *The Daily Princetonian*, August 17, 1942, 1. All-black units with white officers: Richard B. Stolley, ed., *LIFE: A Century of Change*, 219. When the war began, the U.S. marines totally excluded African Americans; the navy used them, essentially, as servants (cooks and stewards); and the army created segregated units. The Red Cross even segregated blood. Later, as the war required more soldiers, more black GIs were "permitted" to see combat duty.

Chapter 7. The Home Front

1. "It had to be clear to Einstein . . . that he was off the atomic invitation list" (Jerome, *The Einstein File*, 40–41). Though he didn't know the details, Einstein was surely aware that a huge government project was under way—"virtually all nuclear physicists [in the

United States], as well as countless scientists from other disciplines had vanished, their addresses unknown [and] Einstein was probably able to form a picture of the state of affairs" (Albrecht Fölsing, *Albert Einstein,* 718).

2. Several war-related organizations: These included Spanish Refugee Relief, the Joint Anti-Fascist Refugee Committee, Russian War Relief, the Committee to Aid Spanish Democracy, the Scientific and Cultural Conference for World Peace, the National Council of American-Soviet Friendship, and the World Congress against War.

3. Einstein's fund-raising letter for the National Council of American-Soviet Friendship, April 7, 1943: Einstein Archives, Box 79, 54-752; rally for American-Soviet friendship, November 14, 1945: FBI's Einstein dossier, section 3, 461.

4. Worked as consultant for the U.S. Navy: From June 18, 1943, to October 15, 1944, Einstein sent navy lieutenant Stephen Brunnauer regular, detailed reports—some handwritten, others typed, including his hand-drawn diagrams—on problems relating to high explosives. Brunnauer later reported that weapons tests had confirmed Einstein's solutions to be completely accurate; Lt. Stephen Brunnauer, "Einstein and the U.S. Navy," cited by Fölsing, *Albert Einstein,* 715. (Brunnauer regularly brought Einstein "problems such as the optimal detonation of torpedoes. His solutions . . . were accurate.") Einstein's letters to Lt. Brunnauer were declassified by the navy in 1979. Copies can be obtained from the National Archives. Fund-raising effort for War Bonds: Sayen, *Einstein in America,* 149.

5. Details of Detroit riot from "Rearview Mirror" report, Vivian M. Bauch and Patricia Azcharias, *The Detroit News,* on-line, 2004; "leadership from left-wing African American workers": Michael Goldfield, *The Color of Politics,* 215–217.

6. Vivian M. Bauch and Patricia Azcharias, *The Detroit News,* "Rearview Mirror" series, on-line, 2004.

7. Ibid.: "only blacks—17 of them—were killed by police. . . . Thurgood Marshall, then with the NAACP, assailed the city's handling of the riot. He charged that police unfairly targeted blacks while turning their backs on white atrocities. He said 85 percent of those arrested were black, while whites overturned and burned cars in front of the Roxy Theater with impunity while police watched.'This weak-kneed policy of the police commissioner coupled with the anti-

Negro attitude of many members of the force helped to make a riot inevitable,' Marshall said."

8. Einstein signing statement to Roosevelt on preventing more Detroit riots: [New York newspaper] *PM,* August 7, 1943, cited in the FBI's Einstein file, section 3, 299; HUAC finding Detroit riots "Communist-inspired": David Caute, *The Great Fear,* 167.

9. "I am a member of the NAACP," Einstein wrote to Murray Gitlin on June 24, 1942; Einstein Archives, Box 80, 55-133–135. Endorsement of Campaign to Save Sam Buckhannon: Einstein's FBI dossier, section 8. Sponsorship of a new NAACP Defense Fund, Einstein's FBI dossier, section 3, 298; and letter in *PM,* August 7, 1943, to Roosevelt on preventing more Detroit riots: section 3, 299. Also on the Buckhannon case: *New York Amsterdam News,* October 2, 1943, 3.

10. Einstein File, Correlation Report, 1153.

11. Hoover's early file on "Radicalism and Sedition among Negroes" and his citing of Randolph's *Messenger* and Du Bois's *Crisis:* David Levering Lewis, *W.E.B. Du Bois,* 6–7. "Negro Activities" file and attack on Garvey: Athan Theoharis and John Stuart Cox, *The Boss,* 57. Theoharis and Cox cite Lowenthal (*The FBI,* 90), as well as Robert Hill, "'The Foremost Radical among His Race': Marcus Garvey and the Black Scare, 1918–1921," *Prologue* (Winter 1984) and a 1984 paper by Theodore Kornweibel Jr., "The FBI and Black America, 1917–1922."

12. Hoover's racism: Richard Gid Powers, *Secrecy and Power,* 324–331; a political moderate: When Fred Jerome, in 1999, asked the FBI's public relations man Rex Tomb to recommend someone who might give the FBI "a fair shake," Tomb, without hesitating, said, "call Richard Gid Powers," and then recited Powers's phone number. Hoover quashed an FBI probe into the . . . bombing: *New York Times,* July 13, 1997, IV:16. See also *New York Times* articles of May 22, 2000, 18; and April 13, 2001, 12.

13. African American "agents" in the FBI: Powers, *Secrecy and Power,* 323. The sixth black agent, highly skilled veteran detective James Amos, was kept in the office supervising weapons inventory in the Bureau's New York office until his death in December 1953; see Kenneth O'Reilly *"Racial Matters,"* 29–30.

14. Edward R. Murrow on CBS radio, August 12, 1945, three days after the bombing of Nagasaki. Growing number of critics: *Catholic*

World (September 1945) and another leading Catholic magazine, *America* (also September), said the Japanese had been prepared to surrender before the bombing. Other religious critics included the Commission on the Relation of the Church to the War in the Light of the Christian Faith, Federal Council of the Churches of Christ in America, March 1946. *The Christian Century*, August 29, 1945, said the bombing "placed our nation in an indefensible moral position."

15. Anti-Japanese comments, racist editorial cartoons: the *Philadelphia Inquirer* denounced "the whining, whimpering, complaining Japs." Ninety-eight percent of the white press supported the bombing: Gar Alperovitz, *The Decision to Use the Atomic Bomb*, 427–428 and Paul Boyer, *By the Bomb's Early Light*, 12–19. Media support for the atom bombing included the leftist press, such as the *New Republic*, the *Nation*, and the Communist party's *Daily Worker*.

16. Criticism from the black press: *Crisis* editorial by Roy Wilkins, September 1945, 249, cited by Boyer, *By the Bomb's Early Light*, 199. *Chicago Defender* articles by Hughes, White, and Du Bois: August 18, September 8, and September 15, respectively. *Washington Afro-American*: August 18, 1945.

Chapter 8. Civil Rights Activist

1. "Double-V": Emily Wax, "Blacks' Other WWII Battle," *Newsday*, November 8, 1998. See also David Williams, *Hit Hard*. "I was in the 92nd Division": Howard "Stretch" Johnson, born in 1915, received two Purple Hearts for injuries suffered while fighting with the 92nd Division; see Jennings and Brewster, *The Century*, 286–287.

2. "Bilbo Urges Mississippi Men to Employ 'Any Means' to Bar Negroes from Voting," headline in *New York Times*, June 23, 1946.

3. Wave of lynching: "Between June 1945 and September 1946, fifty-six blacks were killed in a re-inaugurated reign of terror"; Duberman, *Paul Robeson*, 305; see also David Montgomery, introduction to *The Cold War and the University*, ed. Andre Schiffrin, and Ralph Ginzburg, *100 Years of Lynching*.

4. The Columbia, Tennessee, story: Juan Williams, *Thurgood Marshall*, 131–142.

5. "closer to German storm troopers": Thurgood Marshall quoted in *New York Times*, March 2, 1946, 26. For more on the events in Co-

lumbia and Marshall's role—Marshall himself narrowly escaped from a lynch mob (including local police) that nearly succeeded in murdering the future Supreme Court justice—see Williams, *Thurgood Marshall,* 131–142.

6. Vice chairman of the Democratic National Committee, Oscar W. Ewing, reported on the rise of the Klan in the Midwest: *New York Times,* February 3, 1946, 32.

7. The Committee for Justice in Columbia included an array of celebrities such as Mary Mcleod Bethune, Col. Roy Carlson (Carlson's Marine Raiders), Marshall Field, Oscar Hamnmerstein II, Helen Hayes, Sidney Hillman, Langston Hughes, Harold Ickes, Herbert Lehman, Sinclair Lewis, Joe Louis, Henry Luce, Adam Clayton Powell Jr., A. Phillip Randolph, Artie Shaw, and David O. Selznick; see Gail Williams O'Brien, *The Color of Law: Race, Violence, and Justice in the Post–World War II South,* 263, n. 86.

8. Anton Reiser (nom de plume of Einstein's son-in-law Rudolf Kayser), *Albert Einstein, A Biographical Portrait,* 187. Einstein actually called the honorary degrees "ostentatious rolls."

9. Horace Mann Bond, *Education for Freedom,* chap. 1, cited on Lincoln University's Website.

10. "in a worthwhile cause . . . I do not intend to be quiet": *Baltimore Afro-American,* May 11, 1946, 1.

11. Robert Byrd's letter to Bilbo: Graham Smith, *When Jim Crow Met John Bull.* Byrd, like Bilbo, was then a KKK member. But Senator Byrd, an outspoken critic of the Bush administration's war in Iraq, has radically shifted his views on race: "I will not dispute the quote, though I consider it deplorable," he has said. "I am ashamed to be associated with such despicable sentiments. Becoming involved with the KKK was the most egregious mistake I have ever made. Upon introspection, I find the entire episode difficult to understand. The only conclusion I can draw for myself is that I was sorely afflicted by a dangerous tunnel vision, the kind of tunnel vision that, I fear, leads young people today to join gangs or hate groups." Associated Press, Charleston, W.Va., September 4, 1999.

12. Douglas Starr, *Blood,* 96–99, 108–109, 169–170.

13. *The Lincolnian* 17, no. 4 (June 4, 1946).

14. For Dr. Bond's letter to Einstein, we are grateful to Susan Pevar, archivist, Langston Hughes Memorial Library at Lincoln University

in Pennsylvania. The complete note says: "A few days ago, in the City of Pittsburgh, the mother of one of our students told me—'I was very happy to know that my boy had an opportunity to see Dr. Einstein.' All of us are as grateful as this humble mother." Einstein wore "daddy's mortarboard": Fred Jerome interview with Yvonne Foster Southerland, November 29, 2004.

15. A list of 135 political articles *and* 150 *New York Times* citations just through 1950, prepared by Margaret C. Shields, appears in Paul Arthur Schilpp, ed., *Albert Einstein Philosopher-Scientist,* 691–695. Also Nell Boni, Monique Russ, and Dan H. Lawrence, *Bibliographical Checklist and Index to the Collected Writings of Albert Einstein* lists additional Einstein articles published after 1950. The total number of Einstein's *nonscience* ("general") essays and articles listed in the *Checklist* as published between 1920 and 1955 is actually 307.

16. Front-page stories with photos appeared in the *Philadelphia Tribune,* May 7, 1946, and *Baltimore Afro-American,* May 11, 1946. The *New York Age, New York Amsterdam News* ("Einstein: Race Problem a Disease of 'White Folks'") and *Pittsburgh Courier* all ran stories, with at least one photo, on May 11, 1946.

17. In successive weeks, African American veterans J. C. Farmer in Bailey, North Carolina, and Macio Snipes, the only black man to vote in his district of Taylor County, Georgia, were shot down by bands of white men. As his mother stood 100 yards away, Farmer was killed by bullets from a posse of twenty to twenty-five "deputies" in eight cars. An hour earlier, while waiting for a bus, he had gotten into a scuffle with a policeman. Snipes was gunned down on the porch of his home by ten white men. NAACP, *30 Years of Lynching in the U.S.,* and Ginzburg, *100 Years of Lynching.*

18. Jerome, *The Einstein File,* chapter 6, "The American Crusade against Lynching."

19. Confrontation with Truman: *Chicago Defender,* September 28, 1946. The *Philadelphia Tribune* of September 24 reported that when asked if he was making a threat, "Robeson told newspapermen he assured the President it was not a threat, merely a statement of fact about the temper of the Negro people."

20. FBI's twelve pages on ACEL begin on section 4, 572. The cited quotation is on 582.

21. The KKK was not put under FBI surveillance until 1964: Stolley, ed., *LIFE: A Century of Change,* 216.

22. Hoover on lynching: Memo from Hoover to Attorney General, September 17, 1946 (FBI File 66–6200–44), cited in Powers, *Secrecy and Power,* 563, n. 30.

23. Besides Robeson, the ACEL delegation that confronted Truman in the Oval Office included Rabbi Irving Miller of the American Jewish Congress and Mrs. Harper Sibley, president of the United Council of Churchwomen and wife of the former president of the U.S. Chamber of Commerce. Other ACEL sponsors included W.E.B. Du Bois, attorney Bartley Crum, actress Mercedes McCambridge, Dr. Joseph L. Johnson, dean of Howard Medical School; Metz T. P. Lorchard, editor-in-chief of the *Chicago Defender;* and several black church leaders; see *New York Times,* September 23, 1946, 16, and Duberman, *Paul Robeson,* 674.

24. Letter to the Urban League, September 16, 1946: Einstein Archives, Princeton, document 57-543.

25. National Committee to Oust Bilbo, activities and sponsors: Gerald Horne, *Communist Front?* 16, 56.

26. Mississippi senator Theodore Bilbo interviewed on *Meet the Press,* August 9, 1946.

27. "Listen Mr. Bilbo" by Bob and Adrienne Claiborne was first recorded in 1956 on an album called *Love Songs of Friends and Foes* (Folkways 2453). We are grateful to Smithsonian Folkways for permission to reprint these lyrics.

28. From 1946 to 1956, the CRC, once called "the most successful Communist front of all time," led the national and international drive to save Willie McGee, the Trenton Six, and several similar campaigns (see chapter 9), and published *We Charge Genocide,* which detailed the lynchings and police shootings of blacks, year by year. The study was submitted as a petition to the United Nations. "Reprinted in many languages and many thousands of copies, this work was an international embarrassment for the U.S. government." At the same time, the group organized protests in defense of communists arrested under the Smith Act and against what the CRC viewed as the main legal arms of McCarthyism—the Communist Control Act, the McCarran Act, and congressional investigating committees such as

HUAC. (Gerald Horne in Buhle et al., eds., *Encyclopedia of the American Left*, 134–135.)

Chapter 9. From World War to Cold War

1. "rejuvenated": Wittner, cited in Zinn, *Twentieth Century*, 127. Increases in GNP and corporate profits: Goldfield, *The Color of Politics*, 232–234. More than 40 percent of world income: Jennings and Brewster, *The Century*, 287.

2. Jennings and Brewster, *The Century*, 287.

3. Jump in housing starts: Ibid., 285.

4. Zinn, *Twentieth Century*, 127.

5. "Laid off out of seniority": Goldfield, *The Color of Politics*, 237, cites such interracial, left-led unions as the electrical workers (UE), the Packinghouse Workers and the Farm Equipment Workers in Chicago, Local 600 of the United Auto Workers in Detroit, and a few others led by "the racially progressive Communist Party" (CP).

6. "near the edge of mortal crimes": Martin J. Sherwin, *A World Destroyed*, 110, and Clark, *Einstein, The Life and Times*, 697–701.

7. According to Golden's subsequent memo, Einstein said, "It pains him to see the development of a spirit of militarism in the United States," and compared Americans to the German people at the time of the Kaiser. Einstein added: "Americans are beginning to feel that the only way to avoid war is through a Pax Americana, a benevolent world domination by the United States," and warned that "history shows this to be impossible and the certain precursor of war and grief." But Golden was not swayed from his administration's policy; see Golden's memo to Secretary of State Marshall, June 9, 1947, in *Foreign Relations of the U.S., 1945–1950*, 1: 487–489. In April 2001, fifty-five years later, Golden gave the *New York Times* a different version of that interview: "Einstein said it was essential that a world army be created, under the leadership of the U.S. Unless this was done there would be an atomic war in the next 10 years" (*New York Times*, May 1, 2001). At ninety-one (Golden's age when he spoke with the *Times*), a man may be entitled to a less-than-accurate memory. But in this case, Golden's original memo leaves no doubt that Einstein specifically denounced the trend toward U.S. militarism.

8. Soviet scientists' criticism: *Moscow New Times*, November 26,

1947; Einstein's response, *Bulletin of Atomic Scientists* (February 1948); reprinted in *Ideas and Opinions,* 146–165. Einstein wrote a long response. The Soviet scientists had called Einstein's suggestion for world government "a screen for an offensive [by Western powers] against [socialist] nations." Einstein acknowledged the repeated military and political attacks by Western nations that "Russia has suffered . . . during the last three decades," but argued that, to avoid nuclear destruction, both sides had no choice. "There is no other possible way [but world government] of eliminating the most terrible danger which has ever threatened man."

9. "War is won . . . peace is not": *New York Times,* December 11, 1945; full text of speech in Nathan and Norden, *Einstein on Peace,* 355–356.

10. Einstein's support for national liberation movements: letter to Max Yergan, president of the Council on African Affairs (Robeson and Du Bois were chairman and vice chairman, respectively); see FBI's Einstein dossier, section 8. Also see *Daily Worker,* April 21, 1947, 4.

11. "astounding comeback": Zinn, *Twentieth Century,* 127–128.

12. Wilson's comment, ibid., 127; and blaming the Russians, 128.

13. God gave America the bomb: Churchill cited in Jennings and Brewster, *The Century,* 295.

14. "Your courageous intervention": Einstein letter to Wallace, September 18, 1946 (two days before Wallace was fired). Wallace Archives, University of Iowa, cited in John C. Culver and John Hyde, *American Dreamer: A Life of Henry A. Wallace,* 418.

15. UPI photo of Einstein with Henry Wallace, Frank Kingdon, and Paul Robeson appeared in (among other papers) the *Chicago Star* on November 4, 1947, 2. It is cited in the FBI's Einstein File, section 8, 83–85.

16. Support for Wallace: Wallace Archives, University of Iowa, Culver and Hyde, *American Dreamer,* 418.

17. "in large measure responsible": Doris Miller and Marion Nowak, quoted in Zinn, *Twentieth Century,* 131. Truman's personal dislike for Hoover has been widely cited. He reportedly also feared the FBI would become another Gestapo. At first he asked Congress to give the Civil Service Commission primary jurisdiction for loyalty investigations, but Congress balked, voting to give most power to the

FBI. In November 1947, Truman agreed to give Hoover's Bureau *all* authority for investigating government employees.

18. Truman's loyalty program: Miller and Nowak, cited in Zinn, *Twentieth Century,* 131.

19. Ellen Schrecker (*Many Are the Crimes*) makes a convincing case that the primary target of McCarthy-Hooverism was the left-radical influence in labor unions. Even in Hollywood, the Red scare was most welcomed by the big studios because it helped them "to get rid of militant trade unionists"; see John Howard Lawson, *Film in the Battle of Ideas,* 13.

20. Jerome, *The Einstein File.*

21. ACLU collaboration with the FBI and Ernst's secret insider work for Hoover: Donner, *The Age of Surveillance,* 146–147. Also, Ernst, "Why I No Longer Fear the FBI," *Reader's Digest* (December 1950).

22. Gerald Horne, *Black and Red,* 209. Ben Bell of the Chicago branch informed Roy Wilkins that he had "checked with the FBI."

23. "associate activity on interracial matters with disloyalty": Goldfield, *The Color of Politics,* 270–271.

24. Schrecker, *Many Are the Crimes,* 282.

25. "cracked down": Ibid., Schrecker, 391; vulnerable to red-baiting: David Levering Lewis, *W.E.B. Du Bois,* 526.

26. Schrecker, *Many Are the Crimes,* 393.

27. Robert Harris, *The Nation,* March 3, 1951.

28. CP's role in defense cases: Robin D. G. Kelley, in *Encarta Africana,* on-line newsletter. See also Goldfield, *The Color of Politics,* 192–193.

29. "No way can the Negro man win": Rosalie McGee in "Supreme Court Bars Vital Evidence: State's Case against McGee Based on Blackmail, Perjury," *Freedom* 1, no. 4 (April 1951), 1.

30. Jessica Mitford, *A Fine Old Conflict,* 161.

31. "damn New York Communists": *New York Times,* July 27, 1950, 37.

32. Call for lynching: *Jackson Daily News,* cited in Mitford, *A Fine Old Conflict,* 179.

33. Willie McGee's last letter to his wife; Jessica Mitford, "Defending Willie McGee": Albert Fried, *Communism in America,* 377–378.

34. Black women's groups: Robin D. G. Kelley in *Encarta Africana*, on-line newsletter.

35. Ingram campaign: Horne, *Communist Front?* 205–212.

36. *Saturday Review of Literature*, November 11, 1947.

37. To anyone who would listen, the CRC argued, "The defense of the Communist Party is the first line in the defense of civil liberties for everyone": Gerald Horne's essay on the CRC in Buhl, Buhl, and Georgakas, eds., *Encyclopedia of the American Left*, 134.

38. "become incomprehensible": tape-recorded address to the Decalogue Society of Chicago, February 20, 1954, in *Ideas and Opinions*, 37–39, and Nathan and Norden, *Einstein on Peace*, 600–601.

39. Alice Calaprice, *Dear Professor Einstein*.

40. SCEF and Highlander managed to survive: Schrecker, *Many Are the Crimes*, 391. Rosa Parks at Highlander: Fred Powledge, *Free at Last?* 74–75

41. Einstein, "Message to the Southwide Conference on Discrimination in Higher Education." (See Part II, Document I.)

42. In handcuffs: "The venerated Du Bois . . . handcuffed, fingerprinted, bailed, and remanded for trial": Caute, *The Great Fear*, 176. The PIC as "antinuclear, anti–Cold War": Robin D. G. Kelley in Buhl, Buhl, and Georgakas, eds., *Encyclopedia of the American Left*, 204.

43. "fixed Marcantonio with a long look": Shirley Graham Du Bois, *Du Bois: A Pictorial Biography*. Identical report from Du Bois's close friend and associate, Herbert Aptheker: interview with Fred Jerome in Aptheker's home in San Jose, California, February 19, 2001.

44. Du Bois correspondence: Reel 66, no. 591, Du Bois Papers, University of Massachusetts at Amherst.

Chapter 10. Einstein and Robeson, II

1. The invitation sent via a mutual friend to avoid interception: Conversations between Fred Jerome and Lloyd Brown, October–November 1998.

2. Assassination attempt on Robeson: Howard Fast, *Peekskill*, 82, and photo facing 96.

3. "half-fascistic": Letter to Henry A. Wallace, January 26, 1949, Einstein Archives. The letter primarily expressed Einstein's concern about the establishment of NATO, which he called "a horror."

4. "Under [State Department officials] Shipley and Knight (and Hoover, who directed many of their moves) . . . the singer Paul Robeson and the writer Howard Fast, were denied a passport": Curt Gentry, *J. Edgar Hoover, the Man and the Secrets,* 409.

5. Black churches received threatening phone calls; also, "But you came in with a great man": Conversations with Brown, October–November 1998.

6. Einstein comment on Israel: Lloyd Brown to Gil Noble on *Like It Is,* February 24, 2002. Other commentators have offered a variety of alternative reasons for Einstein's declining the offer to become Israel's president, including his poor health and lack of political experience.

7. Ibid.

8. Robeson's report on his visit to Einstein appeared in his monthly column, "Here's My Story," *Freedom,* November 1952. Brown's interview: *Like It Is,* February 24, 2002.

9. "annihilation": On Eleanor Roosevelt's first nationwide TV program on February 12, 1950, Einstein warned that the hydrogen bomb could lead to world "annihilation" (*Einstein's Universe,* BBC and WGBH Television, 1997).

10. Einstein's "depression" letters: Indiana letter, Nathan and Norden, *Einstein on Peace,* 538–539; "the dear Americans have assumed the Germans' place" and "one stands by, powerless" (to Queen Elisabeth, January 6, 1951), ibid., 554; "hardly ever felt as alienated" (to Gertrude Warschauer, July 15, 1950), Alice Calaprice, *The Quotable Einstein,* 24.

11. "Saddest of all" (to Queen Elisabeth, January 3, 1952): Nathan and Norden, *Einstein on Peace,* 562.

12. From the British Columbia *Union News,* June 6, 1952, 1, under headline "Paul Robeson Enthralled Thousands at Monster Peace Arch Concert": "It is difficult to find the necessary superlatives to describe the Paul Robeson concert at the Peace Arch on May 18 because the numbers of protesters and the degree of enthusiasm attained, exceeded even the wildest dreams of the District Union sponsors."

13. On June 12, 1953, reporting on Einstein's letter to Brooklyn schoolteacher William Frauenglass, the *New York Times* headline read: "'Refuse to Testify,' Einstein Advises Intellectuals Called in by Congress." And on December 17, describing the testimony of Al

Shadowitz: "Witness, on Einstein Advice, Refuses to Say If He Was Red."

14. "never seen him so cheerful" : Kahler quoted by Sayen, *Einstein in America,* 272; "enemy of America": McCarthy quoted in *New York Times,* June 14, 1953.

15. "to tire in that struggle": tape-recorded address to the Decalogue Society of Chicago, February 20, 1954, in *Ideas and Opinions,* 37–39, and Nathan and Norden, *Einstein on Peace,* 600–601. "I must keep fighting until I'm dying": Robeson at a rally in London's Albert Hall in December 1937 to raise funds for the victims of Franco's fascists; Duberman, *Paul Robeson,* 213.

From Einstein's FBI File

1. The ACEL-related entries are compiled in two FBI reports within the Einstein File—Section 4, the 1952 Correlation Report, beginning on p. 572 (the excerpt with the revealing tag-line is from p. 582) and the Summary Report of August 5, 1953. Some of the items were written after the September 23, 1946, protest in Washington, while others were written earlier, which explains why the tense shifts from past to present and back again. The item describing Robeson's "long record of Communist affiliations" was written seven years after the Washington event, appearing in Section 8, pp. 29–30 of the 1953 Summary Report.

Bibliography

Interviews

Lloyd Banks
Morris Boyd
Lloyd Brown
Harriet Calloway
Consuela Campbell
Eric Craig
Penney Edwards-Carter
Fanny Reeves Floyd
James Floyd
Shirl Gadson
Albert Hinds
Timmy Hinds
Wallace Holland
Joi Morton
Terri Nelson
Henry Pannell
Rod Pannell
Alice Satterfield
Shirley Satterfield
Lena Sawyer
Callie Carraway Sinkler
Yvonne Foster Southerland
Lillie Trotman
Mary Trotman
Evelyn Turner
Mercedes Woods

On Albert Einstein

BOOKS AND ARTICLES

Berlin, Isaiah, et al. *Einstein and Humanism.* Jerusalem: Papers from the Einstein Centennial Symposium, 1979.

Bernstein, Jeremy. *Albert Einstein and the Frontiers of Physics.* New York: Oxford University Press, 1997.

Boni, Nell, Monique Russ, and Dan H. Lawrence. *Bibliographical Checklist and Index to the Collected Writings of Albert Einstein.* Paterson, N.J.: Pageant Books, 1960.

Brunnauer, Lt. Stephen. "Einstein and the US Navy." In *Heterogeneous Catalysis.* Houston: Robert A. Welch Foundation, 1983.

Bryan, Denis. *Einstein, A Life*. New York: John Wiley, 1996.

Bucky, Peter A. *The Private Albert Einstein*. Kansas City, Mo.: Andrews and McMeel, 1992.

Calaprice, Alice. *Dear Professor Einstein: Albert Einstein's Letters to and from Children*. Amherst, N.Y.: Prometheus Books, 2002.

———. *The New Quotable Einstein*. Princeton, N.J.: Princeton University Press, 2005.

———. *The Expanded Quotable Einstein*. Princeton, N.J.: Princeton University Press, 2000.

Clark, Ronald W. *Einstein: The Life and Times*. New York: Avon Books, 1972.

Dukas, Helen, and Banesh Hoffman, eds. *Albert Einstein: The Human Side*. Princeton, N.J.: Princeton University Press, 1979.

Einstein, Albert. "Education and World Peace." *New York Times*, November 24, 1934, 17.

———. *Ideas and Opinions*. Based on *Mein Weltbild*. Introduction by Alan Lightman. New York: Modern Library, 1994.

———. "Message to the Southwide Conference on Discrimination in Higher Education, Atlanta University, 1950." In *Discrimination in Higher Education*. New Orleans: SCEF, 1951.

———. *Out of My Later Years*. New York: Philosophical Library, 1950.

———. "To American Negroes." *The Crisis* 39 (1932), 45.

———. "Why Socialism?" *Monthly Review* 1, no. 1 (May 1949).

———. *The World As I See It* [*Mein Weltbild*]. Amsterdam, 1934. Revised edition edited by Carl Seelig. Zurich: Europa Verlag, 1953.

Fölsing, Albrecht. *Albert Einstein: A Biography*. New York: Viking, 1997.

Frank, Philipp. *Einstein: His Life and Times*. New York: Alfred A. Knopf, 1947.

French, A. P., ed. *Einstein, A Centenary Volume*. Cambridge, Mass.: Harvard University Press, 1979.

Friedman, Alan J., and Carol C. Donley. *Einstein as Myth and Muse*. New York: Cambridge University Press, 1985.

Golden, Fred. "Relativity's Rebel." *Time*, December 31, 1999.

Green, Jim, ed. *Albert Einstein*. Melbourne, Australia: Ocean Press, 2003.

Highfield, Roger, and Paul Carter. *The Private Lives of Albert Einstein.* New York: St. Martin's Press, 1993.

Hoffmann, Banesh, with Helen Dukas. *Albert Einstein, Creator and Rebel.* London: Hart-Davis, MacGibbon, 1973.

Holton, Gerald. *Einstein, History and Other Passions.* New York: Addison Wesley, 1996.

Holton, Gerald, and Yehuda Elkana, eds. *Albert Einstein: Historical and Cultural Perspectives.* From the Einstein Centennial Symposium in Jerusalem. Princeton, N.J.: Princeton University Press, 1982.

Jerome, Fred. "Einstein and Martin Luther King Share Common Enemy, Racism." *Our World News* on-line, January 20, 1998.

———. *The Einstein File: J. Edgar Hoover's Secret War against the World's Most Famous Scientist.* New York: St. Martin's Press, 2002.

———. "Einstein, Race, and the Myth of the Cultural Icon." *ISIS* (December 2004).

Levenson, Thomas. *Einstein in Berlin.* New York: Bantam Books, 2003.

Nathan, Otto, and Heinz Norden, eds. *Einstein on Peace.* New York: Simon and Schuster, 1960.

Pais, Abraham. *Einstein Lived Here.* New York: Oxford University Press, 1994.

———. *Subtle Is the Lord.* New York: Oxford University Press, 1982.

Reiser, Anton [pseud. Rudolf Kayser]. *Albert Einstein, A Biographical Portrait.* New York: A. & C. Boni, 1930.

Rosenkanz, Ze'ev, ed. *Albert through the Looking Glass.* Jerusalem: Jewish National and University Library, 1998.

———. *The Einstein Scrapbook.* Baltimore: Johns Hopkins University Press, 2002.

Sayen, Jamie. *Einstein in America.* New York: Crown Publishing, 1985.

Schilpp, Paul Arthur, ed. *Albert Einstein Philosopher-Scientist.* LaSalle, Ill.: Open Court Publishing, 1949; Library of Living Philosophers, 1970.

Schwartz, Richard Alan. "Einstein and the War Department." *ISIS* (June 1989).

———. "The FBI and Dr. Einstein." *The Nation,* September 3–10, 1983.

Simmons, John. *The Scientific 100.* Secaucus, N.J.: Carol Publishing Group, 1996.

Stachel, John. *Einstein from B to Z.* Boston: Birkhäuser, 2002.

———. "Exploring the Man beyond the Myth, Albert Einstein." *Bostonia* (February 1982).

Stern, Fritz. *Einstein's German World.* Princeton, N.J.: Princeton University Press, 1999.

Vallentin, Antonina. *The Drama of Albert Einstein.* Garden City, N.Y.: Doubleday, 1954.

———. *Einstein: A Biography.* London: Weidenfeld and Nicolson, 1954.

Wallace, Irving. *The Writing of One Novel.* New York: Simon and Schuster, 1968.

Weart, Spencer, and Gertrude Szilard. *Leo Szilard: His Version of the Facts.* Cambridge, Mass.: MIT Press, 1978.

VIDEOTAPES AND DVDS

Aigner, Lucien. "A Day with Einstein, Recollections by Lucien Aigner." 1940 photo session in Princeton. Videotape. Lenox, Mass.: Herbert Wolff, K2 Productions, 1994.

Devine, David, and Richard Mozer. *Einstein: Light to the Power of E(2).* Videotape. Toronto: Devine Productions, 1997.

"Einstein Revealed." *Nova.* Videotape. Boston: WGBH, 1996.

Einstein's Universe. Narrated by Peter Ustinov. Videotape. London: BBC and Boston: WGBH, 1997.

On Paul Robeson

BOOKS AND ARTICLES

Armentrout, Barbara, and Sterling Stuckey. *Paul Robeson's Living Legacy.* Chicago: Columbia College Chicago and Paul Robeson 100th Birthday Committee, 1999.

Brown, Lloyd. *The Young Paul Robeson.* Boulder, Colo.: Westview Press, 1997.

Davis, Lenwood G., ed. *A Paul Robeson Research Guide.* Westport, Conn.: Greenwood Press, 1982.

Duberman, Martin Bauml. *Paul Robeson.* New York: Alfred A. Knopf, 1988.

Foner, Philip, ed. *Paul Robeson Speaks.* New York: Citadel Press, 1978.

Gilliam, Dorothy Butler. *Paul Robeson, All-American.* New York: New Republic Book Company, 1976.

Paul Robeson: Tributes and Selected Writings. Compiled and edited by Roberta Yancy Dent, assisted by Marilyn Robeson and Paul Robeson Jr. New York: Paul Robeson Archives, 1979.

Robeson, Paul. *Here I Stand.* Reprint. Boston: Beacon Press, 1998.

Robeson, Paul Jr. *The Undiscovered Paul Robeson.* New York: John Wiley, 2001.

Robeson, Susan. *The Whole World in His Hands.* Secaucus, N.J.: Citadel Press, 1981.

Stewart, Jeffrey C., ed. *Paul Robeson, Artist and Citizen.* New Brunswick, N.J.: Rutgers University Press, 1998.

VIDEOTAPES AND DVDS

Brown, Lloyd. "Interview." *Like It Is with Gil Noble.* TV program. February 24, 2002.

Noble, Gil. *The Tallest Tree in the Forest. Like It Is* special presentation, 1977.

Videotape. St. Louis: Phoenix Films, Timeless Video release, 1994. Contains Robeson speaking on music at some length.

Paul Robeson: Here I Stand. DVD. Directed by St. Clair Bourne, narrated by Ossie Davis. New York: New York: American Masters and Thirteen/WNET, 1999.

GENERAL

Alexander, Stephan. *"Communazis."* New Haven: Yale University Press, 2000.

Alperovitz, Gar. *The Decision to Use the Atomic Bomb.* New York: Random House, 1995.

Anderson, Marian. *Lord What a Morning.* Madison: University of Wisconsin Press, 1956.

Aptheker, Herbert. *Anti-Racism in U.S. History: The First Two Hundred Years.* Westport, Conn.: Praeger, 1993.

———. *A Documentary History of the Negro People.* New York: Carol Publishing Group, 1993.

Belfrage, Cedric. *The American Inquisition, 1945–1960.* Indianapolis: Bobbs-Merrill, 1973.

Bernstein, Barton J., ed. *Politics and Policies of the Truman Administration.* Chicago: Quadrangle Books, 1970.

Beyerchen, Alan D. *Scientists under Hitler.* New Haven: Yale University Press, 1977.

Bond, Horace Mann. *Education for Freedom.* Lincoln, Pa.: Lincoln University, 1976.

Boyer, Paul. *By the Bomb's Early Light: American Thought and Culture at the Dawn of the Atomic Age.* Chapel Hill: University of North Carolina Press, 1994.

Brecher, Jeremy. *Strike!* San Francisco: Straight Arrow Press (a division of *Rolling Stone*), 1972.

Breese, Gerald. *Glimpses of Princeton Life, 1684–1990.* Princeton, N.J.: Darwin Press, 1991.

Buhle, Mari Jo, Paul Buhle, and Dan Georgakas, eds. *Encyclopedia of the American Left.* Urbana: University of Illinois Press, 1992.

Carroll, Peter N. *The Odyssey of the Abraham Lincoln Brigade: Americans in the Spanish Civil War.* Stanford, Calif.: Stanford University Press, 1994.

Caute, David. *The Fellow-Travelers: Intellectual Friends of Communism.* New Haven: Yale University Press, 1988.

———. *The Great Fear: The Anti-Communist Purge under Truman and Eisenhower.* New York: Simon and Schuster, 1978.

Civil Rights Congress. *We Charge Genocide: The Historic Petition to the United Nations for Relief from a Crime of the United States Government against the Negro People.* New York: International Publishers, 1951.

Chipp, Herschel B. *Picasso's Guernica.* Berkeley: University of California Press, 1988.

Cook, Fred J. *The FBI Nobody Knows.* New York: Macmillan, 1964.

Corson, William R. *The Armies of Ignorance.* New York: Dial Press, 1977.

Crane, Ralph. *Century of Change.* Boston: Little, Brown, 2000.

Culver, John C., and John Hyde. *American Dreamer: A Life of Henry A. Wallace.* New York: W. W. Norton, 2000.

Donovan, Robert J. *Conflict and Crisis: The Presidency of Harry S. Truman, 1945–1948.* New York: W. W. Norton, 1977.

DeLoach, Cartha. *Hoover's FBI.* Washington, D.C.: Regnery Publishers, 1995

Demaris, Ovid. *The Director: An Oral History of J. Edgar Hoover.* New York: Harper's Magazine Press, 1975.

De Toledano, Ralph. *J. Edgar Hoover.* New Rochelle, N.Y.: Arlington House, 1973.

Donner, Frank J. *The Age of Surveillance.* New York: Vintage Books, 1981.

Du Bois, Shirley Graham. *Du Bois: A Pictorial Biography.* Chicago: Johnson Publishing, 1978.

Fariello, Griffin. *Red Scare: Memories of the American Inquisition: An Oral History.* New York: W. W. Norton, 1995.

Fast, Howard. *Peekskill.* New York: Civil Rights Congress, 1951.

Fosl, Catherine. *Subversive Southerner.* New York: Palgrave Macmillan, 2002.

Franklin, John Hope. *From Slavery to Freedom.* 2nd ed. New York: Alfred A. Knopf, 1963.

Franklin, John Hope, and Loren Schweninger. *Runaway Slaves: Rebels on the Plantation.* New York: Oxford University Press, 1999.

Fried, Albert. *Communism in America: A History in Documents.* New York: Columbia University Press, 1997.

Friedrich, Otto. *Before the Deluge: A Portrait of Berlin in the 1920's.* New York: Harper & Row, 1972.

Gellately, Robert. *Backing Hitler: Consent and Coercion in Nazi Germany.* Oxford: Oxford University Press, 2001.

———. *The Gestapo and German Society.* Oxford: Clarendon Press, 1990.

Gentry, Curt. *J. Edgar Hoover, the Man and the Secrets.* New York: Penguin Books, 1991.

Ginzburg, Ralph. *100 Years of Lynching.* Baltimore: Black Classic Press, 1962.

Goldfield, Michael. *The Color of Politics.* New York: New Press, 1997.

Hack, Richard. *Puppetmaster: The Secret Life of J. Edgar Hoover.* Beverly Hills, Calif.: New Millennium Press, 2004.

Hageman, John Frelinghuysen. *History of Princeton and Its Institutions.* Philadelphia: J. B. Lippincott, 1879.

Honey, Michael. "The War within the Confederacy: White Unionists of North Carolina." *Prologue* (Summer 1986).

Hoffman, Peter. *The History of German Resistance, 1933–1945.* Cambridge, Mass.: MIT Press, 1977.

Horne, Gerald. *Black and Red: W.E.B. Du Bois and the Afro-American Response to the Cold War, 1944–1963.* Albany: State University of New York Press, 1986.

———. *Communist Front? The Civil Rights Congress, 1946–1956.* Cranbury, N.J.: Associated University Presses, 1988.

Hudson, Hosea. *Black Worker in the Deep South.* New York: International Publishers, New York, 1972.

James, C.L.R. *Fighting Racism in World War II.* New York: Pathfinder Press, 1980.

Jennings, Peter, and Todd Brewster. *The Century.* New York: Doubleday, 1998.

Keenan, Roger. *The Communist Party and Auto Workers Union.* Bloomington: Indiana University Press, 1980.

Keller, William W. *The Liberals and J. Edgar Hoover.* Princeton, N.J.: Princeton University Press, 1989.

Kelley, Robin D. G. *Hammer and Hoe: Alabama Communists during the Great Depression.* Chapel Hill: University of North Carolina Press, 1990.

Kessler, Ronald. *The FBI.* New York: Pocket Books (Simon and Schuster), 1993.

Lawson, John Howard. *Film in the Battle of Ideas.* New York: Masses and Mainstream, 1953.

Leslie, Stuart. *The Cold War and American Science.* New York: Columbia University Press, 1993.

Lewis, David Levering. *W.E.B. Du Bois: The Fight for Equality and the American Century, 1919–1963.* New York: Henry Holt, 2000.

Lichtenstein, Nelson. *Labor's War at Home.* Cambridge: Cambridge University Press, 1982.

Lipsitz, George. *Rainbow at Midnight: Labor and Culture in the 1940s.* Chicago: University of Illinois Press, 1994.

Lowenthal, Max. *The FBI.* New York: William Sloane, 1950.

Marrin, Albert. *The War for Independence: The Story of the American Revolution.* New York: Atheneum, 1988.

Meier, August, and Elliott Rudwick. *Black Detroit and the Rise of the UAW.* New York: Oxford University Press, 1979.

———. *From Plantation to Ghetto.* New York: Hill and Wang, 1976.

Meyer, Gerald. *Vito Marcantonio: Radical Politician, 1902–1954.* Albany: State University of New York Press, 1989.

Miller, Douglas, and Marion Nowak. *The Fifties: The Way We Really Were.* New York: Doubleday, 1977.

Mitford, Jessica. *A Fine Old Conflict.* New York: Alfred A. Knopf, 1977.

Mitgang, Herbert. *Dangerous Dossiers.* New York: Ballantine Books, 1988.

Naison, Mark. *Communists in Harlem during the Depression.* Chicago: University of Illinois Press, 1983.

Nash, Jay R. *Citizen Hoover.* Chicago: Nelson-Hall, 1972.

Nelson, Steve. *Steve Nelson, American Radical.* Pittsburgh: University of Pittsburgh Press, 1981.

O'Brien, Gail Williams. *The Color of Law: Race, Violence, and Justice in the Post–World War II South.* Chapel Hill: University of North Carolina Press, 1999.

O'Reilly, Kenneth. *Black Americans: The FBI File.* New York: Carroll and Graf Publishers, 1994.

———. *"Racial Matters": The FBI's Secret File on Black America, 1960–1972.* New York: Free Press, 1989.

Painter, Nell Irvin. *The Narrative of Hosea Hudson, His Life as a Negro Communist in the South.* Cambridge: Cambridge University Press, 1979.

Parenti, Christian. *The Soft Cage: Surveillance in America from Slavery to the War on Terror.* New York: Basic Books, 2003.

Pendleton, Leila Amos. *A Narrative of the Negro.* Washington, D.C.: R. L. Pendleton, 1912.

Powers, Richard Gid. *G-Men: Hoover's FBI in American Popular Culture.* Carbondale: Southern Illinois University Press, 1983.

———. *Secrecy and Power: The Life of J. Edgar Hoover.* New York: Free Press, 1987.

Powledge, Fred. *Free at Last? The Civil Rights Movement and the People Who Made It.* Boston: Little, Brown, 1991.

Preis, Art. *Labor's Giant Step: Twenty Years of the CIO.* New York: Pioneer, 1964.

Rhodes, Richard. *Dark Sun: The Making of the Hydrogen Bomb.* New York: Simon and Schuster, 1995.

———. *The Making of the Atomic Bomb.* New York: Simon and Schuster, 1986.

Robbins, Natalie. *Alien Ink.* New York: William Morrow, 1992.

Robinson, Cedric. *Black Marxism: The Making of the Black Radical Tradition.* London: Zed Books, 1983.

Schiffrin, Andre, ed. *The Cold War and the University*. New York: New Press, 1977.
Schott, Joseph. *No Left Turns*. New York: Praeger, 1972.
Schrecker, Ellen W. *The Age of McCarthyism*. New York: Bedford Books of St. Martin's Press, 1994.
———. *Many Are the Crimes, McCarthyism in America*. New York: Little, Brown, 1998.
———. *No Ivory Tower*. New York: Oxford University Press, 1986.
Schultz, Bud, and Ruth Schultz. *It Did Happen Here: Recollections of Political Repression in America*. Berkeley: University of California Press, 1989.
Schwartz, Richard Alan. *Cold War Culture*. New York: Facts on File, 1997.
———. *The Cold War Reference Guide*. Jefferson, N.C.: McFarland, 1997.
Sherwin, Martin J. *A World Destroyed: The Atomic Bomb and the Grand Alliance*. New York: Alfred A. Knopf, 1975.
Shirer, William. *The Rise and Fall of the Third Reich*. New York: Simon and Schuster, 1959.
Simpson, Christopher. *Blowback: America's Recruitment of Nazis and Its Effects on the Cold War*. New York: Weidenfeld and Nicolsen, 1988.
Smith, Graham. *When Jim Crow Met John Bull*. New York: St. Martin's Press, 1988.
Starr, Douglas. *Blood: An Epic History of Medicine and Commerce*. New York: St. Martin's Press, 1988.
Stern, Fritz. *Dream and Delusions: The Drama of German History*. New York: Alfred A. Knopf, 1987.
Stolley, Richard B., ed. *LIFE: A Century of Change*. Boston: Little, Brown, 2000.
Stone, I. F. *The Truman Era*. New York: Random House, 1953.
Sullivan, William C., with Bill Brown. *The Bureau: My 30 Years in Hoover's FBI*. New York: W. W. Norton, 1979.
Summers, Anthony. *Official and Confidential: The Secret Life of J. Edgar Hoover*. New York: G. P. Putnam's Sons, 1993.
Swearingen, M. Wesley. *FBI Secrets*. Boston: South End Press, 1995.
Theoharis, Athan. *From the Secret Files of J. Edgar Hoover*. Chicago: Ivan R. Dee, 1991.

———. *J. Edgar Hoover, Sex, and Crime.* Chicago: Ivan R. Dee, 1995.

———. *Spying on Americans.* Philadelphia: Temple University Press, 1981.

Theoharis, Athan, and John Stuart Cox. *The Boss.* Philadelphia: Temple University Press, 1988.

Turner, William W. *Hoover's FBI, the Men and the Myth.* Los Angeles: Sherbourne Press, 1971.

———. *Rearview Mirror: Looking Back at the FBI, the CIA and Other Tails.* Granite Bay, CA: Penmarin Books, 2001.

Ungar, Sanford. *The FBI.* Boston: Little, Brown, 1975.

Vidal, Gore. *United States: Essays, 1952–1992.* New York: Random House, 1992. (See especially Essay 91, "The National Security State.")

Washington, Jack. *The Long Journey Home: A Bicentennial History of the Black Community of Princeton, New Jersey, 1776–1976.* Trenton: Africa World Press, 2005.

Welch, Neil, and David W. Marston. *Inside Hoover's FBI.* Garden City, N.Y.: Doubleday, 1984.

Whitehead, Don. *The FBI Story.* New York: Random House, 1956.

Williams, David J. *Hit Hard.* Toronto: Bantam War Book Series, 1983.

Williams, Juan. *Thurgood Marshall, American Revolutionary.* New York: Times Books, 1998.

Wise, David. *The American Police State.* New York: Random House, 1976.

———. "The FBI's Greatest Hits." *Washington Post Magazine,* October 27, 1996.

Wise, David, and Thomas B. Ross. *The Invisible Government.* New York: Random House, 1964.

Wistrich, Robert S. *Antisemitism: The Longest Hatred.* New York: Schocken Books, 1991.

Wittner, Lawrence S. *Rebels against War: The American Peace Movement, 1941–1960.* New York: Columbia University Press, 1969.

Wright, Bruce. *Black Robes, White Justice.* Secaucus, N.J.: Lyle Stuart, 1987.

Zinn, Howard. *Twentieth Century, A People's History.* New York: Harper & Row, 1980.

Index

About the Authors

Fred Jerome is the author of *The Einstein File: J. Edgar Hoover's Secret War against the World's Most Famous Scientist* (2002). A veteran journalist and science writer, his articles and op-ed pieces have appeared in dozens of publications, including *Newsweek, Technology Review,* and the *New York Times.* As a reporter in the South during the early 1960s, he covered the exploding civil rights movement. In 1979, he created the Media Resource Service, a widely acclaimed telephone referral service putting thousands of journalists in touch with scientists. More than thirty thousand scientists volunteered for the MRS, answering media questions in their areas of expertise. More recently, Jerome has taught journalism at Columbia Journalism School, NYU, and numerous other New York–area universities. In 2002, he developed and taught a course at New School University titled "Scientists as Rebels."

A native New Yorker, Rodger Taylor has written articles on city life, early African American New York, and New York's eighteenth-century African Burial Ground in newspapers and magazines, including *New York Newsday.* The Animating Democracy Initiative and the Saint Augustine's Slave Gallery Committee commissioned him in 2003 to write a piece on the Saint Augustine's Slave Gallery, one of the few existing slave galleries in New York City. Taylor is a member of the Saint Augustine's Slave Gallery Committee, the Seneca Village Committee (a nineteenth-century African American community displaced by the building of Central Park), and was part of the National Steering Committee on the African Burial Ground Project. A supervising branch librarian with the New York Public Library, he presently works on the Lower East Side in Manhattan.

CPSIA information can be obtained at www.ICGtesting.com
Printed in the USA
LVOW07s1953280716

498185LV00001BA/44/P